"I am in love with the way this ESG Program focuses on the business... taking care of the business from the perspective of responsibility, social impact and governance factors. Instrumental dimensions for the society we should and can become. This ESG Certification program is structured in a way that the certified professional in me fully understands: Risk Management, Governance, Corporate Social Responsibility, Strategic Decision Making, Culture... A not-to-be-missed certification program!"

Ramsés Gallego, *CISM, CGEIT, Chief Technologist, Cybersecurity, DXC Technology, Barcelona, Spain*

"ESG is more than an ethical imperative—it is a strategic necessity. Responsible governance ensures that innovation serves humanity, privacy is protected, and technology advances with accountability at its core. Recognising this importance, the authors have constructed a comprehensive and structured guide to ESG certification, using real world insights and strategies. ESG certification isn't just a credential—it's a commitment to responsible leadership, sustainable innovation, and ethical governance. In a global context, where trust, transparency, and impact matter, certification sets you apart as a steward of a better future. This study guide is your essential resource as you prepare for the exam."

Valerie Lyons, *PhD, CISSP, Chief Operating Officer, BH Consulting, Greater Dublin*

"Most ESG efforts are about attending to the pillars for the benefit of growing the client base. In this program you achieve a certification after bringing you through lasting and impactful learning for the lifetime of your career. If you have wondered about the value of ESG, this ESG Certification Program will provide you with applicable and comprehensive information on all three pillars."

Judy Hissong, *President, Nesso Strategies, San Diego, California*

ESG Certification Exam Study Guide

ESG impacts businesses globally, and today it is interconnected with all aspects of business, including, but not limited to, human resources, labor issues, risk management, regulatory compliance, establishing strategic direction, and areas impacted by climate change.

Certifications currently available in the marketplace are sustainability-specific rather than an all-encompassing holistic approach. The authors have developed the most comprehensive ESG certification program on the market.

This two-tiered program is designed for professionals who possess a general understanding of ESG principles and concepts, however, desire to increase, solidify, and demonstrate that knowledge and understanding. This Study Guide will assist candidates with the overall Certification program process – from initial application to study materials and sample questions, this Guide is designed to prepare the candidate to take the ESGCertification™ examination.

ESG Certification Exam Study Guide

Al Marcella and Madeline Parisi

CRC Press
Taylor & Francis Group
Boca Raton London New York

CRC Press is an imprint of the
Taylor & Francis Group, an **informa** business

Designed cover image: Shutterstock Image ID 2372719329

First edition published 2026
by CRC Press
2385 NW Executive Center Drive, Suite 320, Boca Raton FL 33431

and by CRC Press
4 Park Square, Milton Park, Abingdon, Oxon, OX14 4RN

CRC Press is an imprint of Taylor & Francis Group, LLC

ISBN: 978-1-032-67248-9 (hbk)
ISBN: 978-1-032-67250-2 (pbk)
ISBN: 978-1-032-67255-7 (ebk)

DOI: 10.1201/9781032672557

Typeset in Sabon
by codeMantra

To our families and friends,

for their encouragement, support, and love.

And

To the dedicated ESG professionals working to foster

an understanding of the ESG principles

and their global impact.

Contents

Foreword

The reports of the death of Environmental, Social, and Governance (ESG) are greatly exaggerated. (With apologies to Mark Twain.)

The concept of "Environmental, Social, and Governance" (ESG) factors evolved in 2004 in response to values-based investors seeking a method to evaluate nonfinancial risks (e.g., human rights violations and climate change).

ESG is an analytical framework that uses a set of standards to assess an organization's environmental, social, and ethical impact. An ESG framework is an essential tool for informing not only investment decisions but also business operating models.

Over time, the term ESG has been widely misused. "ESG investing" and "ESG company" are misconceptions. In recent years, the politicization of ESG principles has emerged as a contentious issue, particularly in the United States.

But ESG is here to stay, and there's no turning back. ESG concepts have become embedded in everyday life in some notable ways:

- Energy efficiency ratings on appliances and vehicles influence consumer spending.
- Renewable energy options for homes and businesses continue to grow.
- Many companies offer employee wellness programs and promote work-life balance.
- Younger generations (Millennials and Gen Z) value companies that align with their personal values and emphasize social responsibility.
- Local communities are concerned about the impact of corporate activities such as pollution and community engagement.
- Corporate ethics policies and whistleblower protections are the norm.
- Shareholders exercise activism around corporate governance transgressions.

Many professional development offerings tend to focus heavily on the environmental aspect of ESG. What distinguishes the ESGCertification™ Program approach from ESG? The program provides solid foundational knowledge for each pillar while explaining the interrelationships between the three.

For individuals, this certification enhances career prospects as companies increasingly value ESG expertise. Participants are equipped with risk management and governance skills that are applicable across industries.

ESG-certified professionals enhance a company's reputation and support compliance with evolving regulations. Their expertise can inspire innovative ESG strategies that offer a competitive advantage. The holistic perspective provided by this ESG certification enables more balanced corporate decision-making that considers environmental and social impacts alongside financial outcomes.

ESGCertification™ sets standardized knowledge requirements for both Executive and Generalist levels. The certification creates a common language and a framework for ESG implementation.

- The Executive certification reflects the C-suite's responsibility for integrating ESG considerations into high-level decision-making.
- The Generalist certification addresses broader implementation concerns: corporate social responsibility (CSR) initiatives, equity programs, risk management processes, ethical frameworks, and workplace culture.

ESG has moved beyond being an investment framework to becoming integrated into business operations, consumer decision-making, regulatory approaches, and even personal lifestyle choices. Organizations are now expected to demonstrate responsibility across these dimensions, whether they recognize it as "ESG" or simply as requisite ethical behavior.

At this moment, it is crucial to embrace global citizenship. Being a good global citizen means recognizing our interconnectedness with people and environments across the world and acting responsibly with this awareness. I hope people will take advantage of ESGCertification™ to gain skills and confidence they need to lead in this space.

— Caren Shiozaki,
Partner, Fortium Partners, Santa Fe, New Mexico

Preface

By considering ESG Certification, you have taken an important step in your professional development. Congratulations!

Several years ago, the authors recognized that the ESG model – with its three separate yet interrelated pillars – and ESG's global impact are critical to the future of the planet, its people, and its prosperity. We began searching for an existing certification program or process that assured that those working in the ESG space had the means to attest to their ESG knowledge. As we explored the options available, we realized that the current certificate or executive programs were sustainability-focused. Our goal became developing an ESG certification program that encompasses all three pillars of ESG – an important distinction.

We talked with many experts in the ESG field and across market segments about what ESG meant to them, to their industry, and to their clients.

From these conversations, we developed the outline of the ESG Certification Program you see detailed in this Study Guide.

From the many conversations with industry professionals and guiding the development of the ESGCertification™, we identified and sourced a series of resource materials relevant to all three pillars. These resources were then vetted by ESG subject matter experts in each specific ESG pillar. We have done everything possible to ensure that the resources upon which the exam is based are representative of the current standards and ESG practices as of this writing.

Our many years of experience working with certification programs guided us in the development of this companion Study Guide. The objective is to provide you with an easy-to-use, informative guide that prepares you for the testing process, as well as taking the certification assessment. In this Study Guide, you will find test-taking tips, sample exam questions, and vetted answers. There is also a glossary of ESG terms and a practical list of acronyms for reference.

As the program and content authors, we sought a partnership in the certification and delivery arena for the ESG Certification Program. We are pleased to partner with APMG-International, a leader in global accreditation and examination delivery.

We have confidence that this Study Guide will give you the incentive to apply to take the exam and embark on the journey to achieving your ESG Certification.

Examples Are Not Endorsements

This book was prepared based on publicly available resources and contains references to materials that are provided for the reader's convenience. The inclusion of these references is not intended to reflect their importance, nor is it intended to endorse any views expressed or products or services offered by third-party providers. These reference materials may contain the views and recommendations of various subject matter experts and regulatory agencies. Hypertext links, contact addresses, and websites to information created and maintained by other public and private organizations are used and current as of the (submission or manuscript or publication) date.

The opinions expressed in any of these third-party materials included as references or examples do not necessarily reflect the positions or policies of the authors.

The authors do not control or guarantee the accuracy, relevance, timeliness, update, or completeness of any outside, third-party information included in this book. These references should not be construed as an endorsement by the authors or by the publisher. The reader should validate and substantiate any information directly from third-party providers before authorizing the acquisition, implementation, or use of any product or service referenced in this book.

For clarity to the reader, each ESG component – Environment, Social, Governance – is referred to in the Study Guide, and ESGCertification™ Program material, as a pillar.

The words test, exam, and assessment are used interchangeably in the Study Guide and in referenced material.

Examples used, including the chapter "How to Read and Interpret for Understanding," may apply to a specific field or industry, yet they have broader applicability. The candidate is encouraged to read all documents for applicability regardless of their current field or industry.

At the time of writing the ESG Study Guide, resources used were the prevailing documents. The authors chose to use the 2023 version of the Sustainable Development Goals (SDGs) Report 2023: Special Edition. Although a 2024 version was recently released, the 2023 publication provides a robust midpoint assessment of the 2030 Agenda for Sustainable Development.

The resource documents contained in this Study Guide will continue to be used for the ESG Certification exam. For the Study Guide, specific sections of several resource documents have been used and are noted below for reading and sample question development. For the ESG Certification Executive or Generalist assessment, however, knowledge of the entire resource will be a requirement (or is recommended) and used for assessment question development. Once a candidate's application is approved, the full assessment reading requirements list will be provided.

The Study Guide resources are noted in Appendix 2. The following list identifies resources where specific sections have been used for sample questions. If specific document sections are not identified in this list, the entire Study Guide resource is used for the development and inclusion of sample questions and will be used for the ESG Executive or Generalist exams.

Global Reporting Initiative (GRI) Standards: (n.a.), October 5, 2021, "General Disclosures 2021," GRI 2_ General Disclosures 2021.pdf.

For the Study Guide and the online assessment, GRI 1, 2, and 3 are the focus. These three standards are foundational to the GRI framework, ensuring organizations can report on their sustainability process in a structured, transparent, and standardized manner.

United Nations Sustainable Development Goals (SDGs): (n.a.), July 2023, "The Sustainable Development Goals Report 2023, Special Edition" United Nations System, https://unstats.un.org/sdgs/report/2023/The-Sustainable-Development-Goals-Report-2023.pdf.

For the Study Guide, the Sustainable Development Goals (SDGs) Report 2023: Special Edition focuses specifically on Goal 5 – General Equality – and on Goal 7 – Affordable and Clean Energy. For the online assessment, all 17 goals may be used to generate questions.

We encourage you to read and use this Study Guide, complete the sample questions, and apply to become a certified ESG professional! https://apmg-international.com/

Acknowledgments

The authors wish to thank the many contributors who have provided input to this Study Guide and certification program as it transformed from a concept to this final product. We thank you.

The authors wish to recognize and thank the following for their contributions and for so graciously responding to requests for further information.

John F. Boyce, Head-People Development, AMSOIL Inc., Duluth, Minnesota, and Adjunct Faculty, Northwestern University, Evanston, Illinois, for providing insight into the program and book development at the earliest stages of transferring the concept to a Study Guide.

Richard Cartwright, PE, CHMM, EHS-HAZMAT Compliance Management Solutions Advocate for his counsel at the initial stages of discussion.

Judy Hissong, President, Nesso Strategies, San Diego, California, for providing subject matter and resource material review.

Nahla Nabil, Abu Dhabi, UAE, for providing subject matter and resource material review.

The authors also wish to thank the following for their global perspective and input as we moved through the ESG certification and Study Guide development process.

APMG-International and Chief Executive Officer, Richard Pharro, for his direction in working through the scheme, format, and certification mechanism. We are extremely appreciative and excited as we partner with APMG-International for the management of the ESGCertification™ Program.

Armanino Advisory, Kris Russell, Senior Manager, Sustainability Practices, and Partner, Ron Steinkamp for discussion and insights at the earliest stages of the ESGCertification™ Program concept.

The ESG Exchange and Carolynn Chalmers, Chief Strategy Officer, and Peadar Duffy, Chief Executive Officer, for conversations regarding ESG training and instructional development at various levels of maturity.

The authors also thank the anonymous reviewers identified by the Publisher, CRC Press, who initially reviewed our proposed Study Guide and provided positive feedback and comments on the content when the authors initially brought the concept for this book to CRC Press.

Author Biographies

Al Marcella, PhD, CISA, CISM, President of Business Automation Consultants (BAC) LLC, is an internationally recognized public speaker, researcher, IT consultant, and workshop and seminar leader with 46 years of experience in IT audit, risk management, IT security, and assessing internal controls, having authored numerous articles and more than 35 books on various IT, audit, and security-related subjects. Marcella's clients include organizations in financial services, IT, banking, petrochemical, transportation, services industry, public utilities, telecommunications, and departments of government and nonprofits.

Madeline Parisi, M. Ad. Ed, Principal, founded Madeline Parisi & Associates LLC (MPA) in 2013, as an international organization providing business training content and training material, after a distinguished career working with professional associations. MPA assists organizations through the development of business training materials, in-house and virtual training, white-label writing services, and professional certification training and certification exam question development services.

Introduction

Environmental, Social, and Governance (ESG) continues to grow and is more than just an internal policy with checkboxes; it is also a means to validate to management, staff, and clients that individuals assessing the ESG program or assisting management in meeting ESG objectives and compliance have sufficient operating knowledge to perform the assessment or provide advisory services. We recognize the strategic value of ESG stewardship globally and deliver an ESG Certification program to meet the growing and future market demands for ESG professionals.

We spoke with many experts and practitioners in the ESG space and various industries about what ESG meant to them and to their industry. We had detailed discussions about how they had assurance that their staff and consultants had bona fide ESG knowledge and how clients understood ESG implementation and audit. From these conversations, we sourced a series of materials relevant to all three ESG pillars and had them vetted in various formats for relevance. The material in this Study Guide has been designed and written to support ESG professionals on their path to certification.

Certification Level: The ESGCertification™ – Executive recognizes the holder as having successfully completed a professional assessment examination of the ESG framework that assists an organization in identifying, recording, assessing, and disclosing risks to investors, lenders, insurers, and other stakeholders.

Certification Level: The ESGCertification™ – Generalist recognizes the holder as having successfully completed a professional assessment examination, which addresses the fundamental ESG factors instrumental to the successful operation of global market economies and industries.

In this two-tiered program, each certification is designed for professionals who possess a general understanding of ESG principles and concepts, however, desire to increase, solidify, and demonstrate that knowledge and understanding.

We sourced a series of materials on which the ESGCertification Program is based. These materials were independently vetted by ESG subject matter experts for relevance to each pillar. This list was then cross-referenced by an analysis through OpenAI's generative-AI. The final resource documents identified and used as the source for the ESGCertification Program are the result of this two-step review process.

One of the goals of this ESGCertification Program is that ESG be seen and recognized as a global strategy for businesses to succeed and be profitable while doing good and treating our world as good stewards. As ESG continues to evolve and mature, regulations and compliance, governance, and social culture are foremost in our ethical behavior.

This Study Guide is developed to assist you on your path to your ESG certification. The chapters are organized to help you be successful, starting with an application process how-to, moving to a chapter designed to guide you in reading resource material for understanding, and chapters providing sample questions that are the same structure as the actual test.

Chapter 1

Exam Format and Structure

This chapter introduces the ESGCertification™ Program and its significance. The certification serves as a mark of proficiency for individuals who have demonstrated comprehensive knowledge of ESG principles. It reassures companies and clients that an individual possesses the necessary skills and understanding to navigate the complex world of ESG and sustainability.

ESG CERTIFICATION OVERVIEW

The ESG Certification Exam Study Guide serves as a comprehensive resource for individuals preparing for the Environmental, Social, and Governance (ESG) certification exam. The guide is structured to provide essential knowledge, practical insights, and guidance on the concepts and frameworks that define ESG principles, which have become a critical part of modern business operations across industries. The Study Guide is aimed at both Executives and Generalists in the ESG space, offering a thorough understanding of the subject and how to excel in the certification process.

The guide outlines the meticulous approach taken by the authors and external subject matter experts (SMEs) to determine the selection of the documents and resources included in the study materials. These resources are carefully vetted for accuracy and relevance, and each is substantiated with a four-point justification that is included in the resource section of the guide.

The exam administration process is administered by APMG-International, a highly regarded global accreditation body and examination institute. Consult the APMG website as it is the prevailing source of all current information regarding the ESGCertification Program and how to book public exams – https://apmg-international.com/.

The ESG Certification Exam is presented as an assessment designed to evaluate the candidate's ability to integrate ESG practices into corporate strategies and decision-making processes. The chapter defines the exam's goals and objectives, including ensuring that candidates can apply ESG principles in real-world scenarios. The audience for this certification spans

DOI: 10.1201/9781032672557-1

various industries, from finance and governance to environmental organizations and multinational corporations. It is intended for professionals aiming to enhance their expertise in ESG matters and integrate these principles into their work.

After discussing with many experts in the field and various industries about what ESG meant to them and to their industry, and what guiding content they used for themselves, their staff, and third-party consultants, relevant resource material covering all three pillars was sourced and externally vetted. The resources used underscore the importance of comprehending foundational documents such as SASB, ISO 14001, IFRS, SEC Climate Disclosures, Organisation for Economic Cooperation and Development (OECD), ISO 26000, and others, which you will locate in Appendix 2. These resources are integral to the exam, and candidates are encouraged to familiarize themselves with them to ensure a deep understanding of ESG principles on which the assessments are based. The section also details the exam format and content – where questions will come from, including an extensive list of study resources. Successful completion of the exam is defined by meeting the required score thresholds, demonstrating the candidate's knowledge, and understanding of ESG.

The assessment questions incorporate all three ESG pillars – Environment, Social and Governance.

THE ESGCERTIFICATION™ PROCESS

The first step in the process is to determine your appropriate exam level, either the ESG Executive-level or the ESG Generalist-level certification. You may consider starting with one level and pursuing the other level separately. These eligibility requirements will assist in determining the appropriate certification level for you.

Content and materials are located on the APMG-International website under ESGCertification™ – https://apmg-international.com/. The APMG website is the prevailing source of all current information regarding the ESGCertification Program and how to book public exams.

- Register and complete the online application from the APMG website.
- Submit the appropriate fee with your application.
- The exam is administered online, on-demand. Once approved, schedule your exam.
- Prepare for the assessment.
  - Use this Study Guide and the sample exam questions.
  - Read the selected resources – the assessment is based on these documents. You will be directed to specific sections where required otherwise read the resource in its entirety.

    - Take the sample assessment included for your specific certification level.
    - Retake the sample assessment as needed.
- Take the online assessment as scheduled.
- Upon successfully completing the assessment, earn your certification.
    - Download your digital certificate.
    - Receive your digital badge, which you can share on social media.

Exam Fees

The exam fee will be presented in your local or home currency as you process your payment through the APMG website. Fees are subject to change and will reflect the exchange rate at the time of payment.

Retaking the Exam

Candidates who do not successfully pass the assessment will have the opportunity to retake the assessment at a future date. Information regarding the retake process will be made available to those candidates.

MORE ABOUT THE RESOURCES, DOMAINS, AND STRUCTURE

To assist the candidate, the methodology for interpreting the resources provided in the study guide and to provide an approach to confidently answer both the ESG Study Guide sample questions and certification exam is covered in Chapter 3: Reading for Comprehension and Understanding: Principles and Strategies. By utilizing a sample article, you will have the opportunity to apply the principles and strategy to read for understanding as related to taking a multiple-choice exam. To further aid in developing this ability, sample multiple-choice questions are included in Chapter 3 to help apply the methodology.

Discussion with professionals working in the field, ESG SMEs, and research into current regulations and global advances led to the development of the two-level ESG Certification Program – Executive and Generalist.

The ESG Executive Certification focuses on the senior leader responsible for integrating ESG principles into corporate strategy, risk management, and reporting. The study materials and assessment examination have as their foundation four constructs essential to identifying, recording, assessing, and disclosing ESG-related financial risks. The constructs or domains included in the Executive Certification incorporate, but are not limited to, topics such as risk management, governance, strategy, and metrics and targets. The Executive-level professional must possess strategic foresight, regulatory expertise, and leadership skills to navigate sustainability challenges and drive meaningful corporate ESG initiatives and possess the desire to increase and solidify that knowledge and understanding.

The Generalist Certification focuses on mid-level sustainability professionals who may be responsible for implementing, reviewing, and maintaining ESG initiatives within organizations. The domains included in the Generalist Certification include, but are not limited to, topics such as corporate social responsibility (CSR), diversity, equity, inclusion, accessibility, and justice (DEIAJ), risk management, ethics, and workplace culture and responsibility. Generalists act as a bridge between strategy and execution, ensuring that ESG commitments translate into actionable outcomes.

The ESG Executive Certification study documents are based on material developed in part by the Task Force on Climate-Related Financial Disclosures (TCFD). TCFD was created in late 2015 to help identify information needed by the financial sector to appropriately assess and price climate-related risks. Since its development, and concurrently with the release of its 2023 status report on October 12, 2023, the TCFD fulfilled its remit and disbanded. The IFRS Foundation has been requested to take over monitoring the progress of companies' climate-related disclosures.[1]

Additional Certification study materials were designed using internationally recognized sources, which include, but are not limited to, Sustainability Accounting Standards Board (SASB), ISO 14001 and ISO 26000, IFRS, Security and Exchange Commission Climate Disclosures, and the OECD (see Appendix 2 for the full list).

The participant will utilize a self-study approach to prepare for the certification assessment exam. Study materials are freely available to the candidate.

EXECUTIVE ESG CERTIFICATION

The ESG Executive Certification recognizes the holder as having successfully completed a professional assessment examination of the ESG framework that assists an organization in identifying, recording, assessing, and disclosing ESG-related financial risks to investors, lenders, insurers, and other stakeholders.

Audience:

The target audience for the ESG Certification – Executive level includes but is not limited to the following professionals:

- ESG Executive-level professionals

- Board Members, Directors, experienced Executives, and Managers seeking to advance their knowledge of ESG initiatives and build the necessary skills to direct or oversee organizations' ESG programs, enhancing their contributions in and out of the boardroom.
- Chief Audit Executives
- Chief Risk Officers
- Chief Sustainability Officers
- Head of Sustainability Investing
- Chief HR Officers
- Chief DEI Officers

Prerequisites: There are no prerequisites required to take the Executive ESG Certification exam.

Certification Objectives:

1. Identify the metrics and targets used to assess and manage relevant ESG-related risks and opportunities where such information is material.
2. Understand how the company identifies, assesses, and manages ESG-related risks.
3. Evaluate the actual and potential impacts of ESG-related risks and opportunities on the company's businesses, strategy, and financial planning, where such information is material.
4. Define how the organization's governance structure addresses ESG-related risks and opportunities.
5. Assess the company's governance performance related to addressing ESG-related risks and opportunities.
6. Examine the tangible material impacts of ESG-related risks and opportunities on the entirety of the organization, including strategy and financial planning.

EXECUTIVE ESG CERTIFICATION QUESTIONS

The Executive ESG Certification questions are designed using the holistic definition of an ESG Executive as a senior leader responsible for integrating Environmental, Social, and Governance principles into corporate strategy, risk management, and reporting. This role involves engaging stakeholders, ensuring regulatory compliance, and driving sustainable business practices to enhance long-term value creation. The ESG Executive will ultimately possess strategic foresight, regulatory expertise, and leadership acumen to navigate sustainability challenges and drive meaningful corporate ESG initiatives.

Executive ESG Certification Constructs – Description and Domain Content

Risk Management Integration and Disclosure

Questions within this domain examine the process of integrating ESG-related risks into an organization's existing risk management processes and the disclosure of these risk management processes in alignment with international frameworks.

Content may include the following:

- Define fundamental risk management concepts and terms to assure: (a) comprehension of risk management approaches and application; and (b) consistency in the use and application of risk management terminology.
- Discuss the unique characteristics of ESG-related risks that are important to consider when integrating such risks into existing processes.
- Identify sustainability-related risks including understanding how "materiality" influences the risk assessment process.
- Examine the practicalities of integrating ESG-related risks into existing risk management processes.
- Analyze risk management disclosure procedures to provide context for how an organization thinks about and addresses the most significant risks to successfully executing its business objectives and accomplishing its strategy, and provide some examples of how different companies disclose risk management information.

Governance

This domain focuses on the organization's approach to governance around ESG-related risks and opportunities. This domain further reviews recommended governance disclosure approaches that apply to all organizations.

Content may include the following:

- The importance of disclosing governance practices to provide context for investors and other users.
- Describe the board's oversight of ESG-related risks and opportunities.
- Review management's role in assessing and managing ESG-related risks and opportunities.
- Developing governance processes, which cover ESG-related issues, to avoid creating separate processes or duplicating existing governance disclosure requirements.

Strategy

This domain reviews the actual and potential impacts of ESG-related risks and opportunities on the company's businesses, strategy, and financial planning. This domain further reviews recommended disclosure strategies applicable to all organizations.

Domain content may include the following:

- For companies that have made Greenhouse Gas (GHG) emissions reduction commitments or operate in jurisdictions that have made such commitments or have agreed to meet investor expectations regarding their GHG emissions, communicate the organization's plans for transitioning to a low-carbon economy.
- Understand the link between the Environmental, Social, and Governance factors to the overall management and performance of the organization.

Metrics and Targets

The focus of this domain describes developments around ESG-related metrics and users' increasing focus on information describing organizations' plans for implementing and moving ESG policies forward. This domain also examines a suite of cross-industry, ESG-related metric categories that all organizations, regardless of market sector, can disclose.

Domain content may include the following:

- An examination of information on selecting and disclosing metrics, including a review of a set of metrics that all organizations should disclose.
- Provides direction on selecting and disclosing ESG targets as well as details on the role of scenario analysis in determining targets.
- Assess how SG metrics, targets, and information from transition plans provide useful underlying information with which to estimate the actual or potential impact of ESG issues on an organization's performance and position.

ESG GENERALIST CERTIFICATION

The ESG Generalist Certification recognizes the holder as having successfully completed a professional assessment examination, which addresses the five fundamental ESG factors instrumental to the successful operation of global market economies and industries. The ESG Generalist Certification is designed for professionals who possess a general understanding of ESG principles and concepts and desire to increase and solidify that knowledge and understanding. The participant will utilize a self-study approach to prepare for the certification assessment exam. Study materials will be available for purchase by the candidate.

Audience:

The target audience for the ESG Certification – Generalist level includes but is not limited to the following professionals:

- Compliance Officers
- Examiners
- Finance and Accounting Professionals
- DEI Administrators
- HR Professionals
- Internal/External Auditors (financial, operational, IT)
- Regulators
- Risk Management Professionals
- Sustainability Site Technicians
- ESG Site Auditor
- Professionals who desire to gain further insight, knowledge, and understanding of an organization's ESG framework intending to assist its stakeholders in understanding how the organization is managing risks and opportunities related to Environmental, Social, and Governance criteria.

Prerequisites: There are no prerequisites required to take the ESG Generalist Certification exam.

Certification Objectives:

1. Appreciate the rights of others as an ethical commitment to diversity, inclusion, compliance, and CSR.
2. Comprehend the broader concept of CSR.
3. Understand that different cultural approaches exist by geographic region when trying to gauge company culture and employee satisfaction.
4. Describe the role that DEIAJ plays as a part of the broader perspective to an organization's ESG program.
5. Identify transition risks and physical risks and the impact each has on an organization's ability to successfully meet its mission, goals, and objectives.
6. Recognize factors and learn strategies that contribute to a positive work environment for employees and managers.

7. Learn how institutions in society, including government, business, and nonprofits, all participate in devising solutions to environmental issues.
8. Recognize the ethics that organizations display in practice, beginning with moral choices made by management and employees.
9. Summarize the processes for identifying, managing, and assessing financial and non-financial risks associated with ESG initiatives.
10. Verify that the objective of social responsibility is to contribute to sustainable development.
11. Understand the process of developing sustainable organizations for water, land, air, and urban management at the local to global level.

GENERALIST QUESTIONS

The ESG Generalist questions are designed using the understanding that a Generalist in the ESG space is a mid-level sustainability professional responsible for implementing, reviewing, and maintaining ESG initiatives within an organization. They work across departments to ensure compliance with sustainability reporting standards, regulatory requirements, and CSR goals. ESG Generalists are the bridge between strategy and execution, ensuring the company's ESG commitments translate into practical actions and measurable outcomes.

Generalists ESG Certification Constructs – Description and Domain Content

Corporate Social Responsibility

Planet, people, and profit are mutually inclusive!

This domain focuses on the broader concept of CSR. CSR refers to how a company conducts its business and to the belief that businesses have an obligation to society beyond their commitments to their stockholders or investors. In addition to generating profits, companies are expected to have some responsibility to stakeholders such as employees, customers, communities, and the environment. CSR includes corporations being economically responsible, improving labor practices, embracing fair trade, mitigating environmental damage, giving back to the community, and increasing employee satisfaction.

The domains will also review the fiduciary duties of directors to reflect the expectations of corporate stakeholders regarding oversight of corporate affairs. Directors are expected to make inquiries to management to obtain the information necessary to monitor, evaluate, and oversee compliances with the organization's ESG mandates and initiatives.

Domain content may include the following:

- Fraud
  - Anti-Bribery
  - Anti-Corruption
- Leadership Response and Interaction with Shareholders
- ISO 26000 – Social Responsibility
- Organisation for Economic Cooperation and Development (OECD)
- Stakeholder Relations.

Diversity, Equity, Inclusion, Accessibility, and Justice (DEI/DEIAJ)

This domain focuses on the role that DEIAJ has as part of the broader perspective to an organization's ESG program. This domain focuses on areas within DEIAJ that will help an organization position itself in the marketplace, in general, and with investors and shareholders where social factors are heavily weighted. Long-term, DEIAJ can be used as a path to sustainability.

Domain content may include the following:

- Anti-Harassment
- Child Labor
- Cultural Awareness
- Respect for Human Rights
- Local Community Relations
- Environmental Justice.

ESG Risk Management

Managing risk is about the application of policies and procedures to the tasks of identifying, analyzing, and assessing risks, determining the degree of exposure to risk that organizations are willing to accept, and taking appropriate steps to avoid litigation, loss of reputation, or injury.

This domain addresses the processes for identifying, managing, and assessing financial and non-financial risks associated with ESG initiatives. Also discussed will be how those processes are integrated into an organization's overall risk management program. Integrating identified ESG risks into a broader enterprise risk management (ERM) process will assist organizations in identifying, assessing, and managing risks and will contribute directly to improving the truthfulness of corporate disclosures.

Effective risk management will also allow an organization to be better prepared to address both the opportunities and risks that are associated with ESG initiatives.

Domain content may include the following:

- Analyzing Risk
- Traditional Risk Management Strategies

Ethics

This domain focuses on the ethical behavior of Executives and corporate boards, across all market sectors, is getting prime-time attention. Environmental, Social and Governance (ESG) are the three core factors for measuring the sustainability, responsibility, and ethical impact of an investment. ESG has brought ethical business practices to the forefront by creating metrics for emphasizing positive financial outcomes. However, this drive for reporting positive outcomes may lead some organizations to undertake unethical business practices and engage in reporting misinformation, i.e., "ethics-washing."

The continuing influence and impact that personal and professional ethics exerts upon individuals, organizations and society – as well as the factors that influence organizational ethics and affects businesses globally – are examined.

This domain provides insight into designing and implementing processes aimed at creating, assessing, and guiding organizational ethics aimed at preserving and sustaining planet Earth.

Domain content may include the following:

- Discussions on what ethics is?
- What is Ethical Leadership?
- Relationship Between Ethics and Integrity
- Ethical Decision-Making in Complex Situations
- Corporate Transparency and Accountability

Workplace Culture and Responsibility

This domain focuses on how the beliefs and values that have existed in an organization over time create a workplace "culture." Workplace culture is the personality of the organization, the way in which an organization accomplishes its objectives, which is based on: its expectations, values, experiences, mission, philosophy, and vision. Culture is expressed in the organization's self-image, inner workings, interactions with the outside world, and future expectations, and varies globally.

Also discussed will be how the type of work and the work environment itself can directly contribute to the health of workers. This domain further examines how explicit behaviors and norms of leadership can foster a sense of belonging, value, and meaningful contribution among all workers.

Domain content may include the following:

- Psychosocial Work Environment
  - Freedom of Association
  - Equality – Gender, Race, Pay Ratios
  - Personal health resources
  - Labor Standards in the Supply Chain and third-party contractors
- Enterprise Involvement in The Community
  - Social Enterprise Partnering
- Physical Work Environment
  - Working Conditions

NOTE

1 Task Force on Climate-Related Financial Disclosures | TCFD, Retrieved March 14, 2025. https://assets.bbhub.io/company/sites/60/2021/10/FINAL-2017-TCFD-Report.pdf

Chapter 2

Successfully Passing Multiple-Choice Exams

Principles and Strategies

> During a test, people look up for inspiration, down in desperation, and left and right for information.
>
> Unknown

INTRODUCTION

Multiple-choice questions are fairly straightforward. There are usually four answers to choose from for each question, and the test is designed so that you can only pick one answer for each question.[1]

Multiple choice test questions, also known as items, are a globally accepted means of assessing a test candidate's ability, knowledge, and understanding of a specific content area (e.g., the pillars of ESG).

The material in this chapter is presented to assist the test taker in better preparing to successfully pass a professional certification Selected-Response Assessment[2] exam.

A CAVEAT

This document does not replace a proactive approach to studying and preparing for the exam before sitting for and taking the exam.

No "strategies" document or "tip sheet" will replace solid preparation and studying. You must know the material and have prepared well. Utilize the material in this chapter *as an aid* in your preparation process, *not as a substitute* for the requisite study required for successfully passing the examination.

With solid preparation and understanding of the exam material and armed with solid principles and strategies regarding the science and art of "deciphering" a multiple-choice question, you will be better prepared for success when you take the exam.

DOI: 10.1201/9781032672557-2

COMPONENTS OF MULTIPLE-CHOICE QUESTIONS

Multiple-choice questions are usually either incomplete statements followed by possible ways the statements may be completed, or they are questions followed by possible answers.

A multiple-choice question consists of *two* parts:

1. A problem or question, known as the stem
2. A list of suggested solutions, known as alternatives.

The alternatives are further broken down and consist of:

- One correct or best alternative, which is the answer
- Incorrect or inferior alternatives are known as distractors (see Figure 2.1)

Let's take a look at the following example.

Topic: Understanding Social Performance in ESG

The Stem

According to the article, "Advancing the 'S' in ESG: A Primer for CFOs"[3] which of the following is a key responsibility of CFOs in advancing social performance within their organizations?

Stem (Question Prompt)	According to the article, Advancing the "S" in ESG: A Primer for CFOs[2], which of the following is a key responsibility of CFOs in advancing social performance within their organizations?
Options (Answer Choices)	a. Ensuring that ESG considerations are integrated into enterprise risk management and reporting requirements. b. Limiting financial reporting to traditional metrics and avoiding integration with non-financial disclosures. c. Focusing solely on regulatory compliance without engaging with investors or stakeholders. d. Ignoring social performance as it has minimal impact on financial outcomes.
Correct Answer	✓ a) Ensuring that ESG considerations are integrated into enterprise risk management and reporting requirements.
Distractors (Incorrect Choices)	✗ b) Incorrect because the document emphasizes integrating financial and non-financial data for holistic reporting. ✗ c) Incorrect because engaging with investors, regulators, and stakeholders is a key role of CFOs in advancing ESG. ✗ d) Incorrect because the document highlights that social performance directly affects business resilience, reputation, and profitability.

Figure 2.1 Elements of a Multiple-Choice Question.[4]

The Alternatives

A. Ensuring that ESG considerations are integrated into enterprise risk management and reporting requirements.
B. Limiting financial reporting to traditional metrics and avoiding integration with non-financial disclosures.
C. Focusing solely on regulatory compliance without engaging with investors or stakeholders.
D. Ignoring social performance as it has minimal impact on financial outcomes.

The Distractors

The following choices are incorrect because:

B. The document emphasizes integrating financial and non-financial data for holistic reporting.
C. Engaging with investors, regulators, and stakeholders is a key role of CFOs in advancing ESG.
D. The document highlights that social performance directly affects business resilience, reputation, and profitability.

The CORRECT Answer

A. Ensuring that ESG considerations are integrated into enterprise risk management and reporting requirements.

Substantiation

(n.a.), March 7, 2023, "Advancing the S in ESG: A Primer for CFOs," World Business Council for Sustainable Development (WBCSD), [Introduction, page 2] www.wbcsd.org/resources/advancing-the-s-in-esg.

All well-designed and formatted multiple-choice questions will exhibit this architectural construction.

The remainder of this chapter will provide insights, examples, principles, and tips on how to interpret, disassemble, analyze, read, and answer Selected-Response Assessment questions (i.e., multiple-choice questions).

Educated Guessing Strategies in Multiple-Choice Exams

Educated guessing strategies refer to techniques that test takers use to logically eliminate incorrect answer choices and make a more informed selection when unsure of the correct answer in multiple-choice exams.

These strategies help increase the probability of selecting the right answer by applying logical reasoning, patterns, or common tendencies found in test design.

TYPES OF EDUCATED GUESSING STRATEGIES

1. Frequency of Occurrence Strategy
   - Definition: This strategy suggests that answer choices that appear frequently in the exam or a specific subject area are more likely to be correct. Test makers often include common facts, principles, or concepts that repeat across multiple questions.
   - Why It Works: If a term, concept, or numerical value is referenced multiple times in the test, it is likely to be significant and could indicate a correct answer.
2. Highest and Lowest Numbers Strategy
   - Definition: When faced with numerical answer choices, extreme values (either the highest or lowest) are often less likely to be correct, while moderate values tend to be the right choice.
   - Why It Works: Test makers generally avoid using extreme numbers as correct answers unless the question is explicitly testing an outlier or record-breaking statistic.
3. Umbrella Strategy
   - Definition: This strategy involves selecting the answer that is the most comprehensive or general, meaning it includes elements of multiple other answer choices.
   - Why It Works: Many correct answers in multiple-choice exams are broad statements that encompass a variety of concepts rather than being overly specific.

Using the document *Advancing the "S" in ESG: A Primer for CFOs*,[5] the following multiple-choice questions demonstrate the application of each Educated Guessing Strategy defined above.

1. Frequency of Occurrence Strategy

   Which of the following best describes the role of CFOs in advancing the "S" (social) in ESG?

   a. CFOs should focus solely on financial metrics and avoid non-financial reporting.
   b. CFOs play a crucial role in measuring social outcomes and integrating ESG into financial decision-making.
   c. The "S" in ESG is primarily a concern for marketing and public relations teams.

d. Social factors in ESG are unregulated and do not impact corporate performance.

Correct Answer

b. CFOs play a crucial role in measuring social outcomes and integrating ESG into financial decision-making.

Why?

The phrase "CFOs play a role in measuring social outcomes" appears frequently in the provided document, reinforcing its importance.

2. Highest and Lowest Numbers Strategy

According to the report, how many key focus areas should CFOs consider when evaluating "S" outcomes in ESG?

a. 2
b. 4
c. 6
d. 12

Correct Answer

c. 6

Why?

While the lowest (2) and highest (12) numbers seem unlikely, a moderate number like six aligns with typical structured frameworks and is correct based on the report.

3. Umbrella Strategy

Which of the following best represents the broad approach CFOs should take when incorporating social responsibility into ESG practices?

a. Ensuring fair wages for direct employees.
b. Implementing safety measures for workers in company-owned facilities.
c. Engaging with stakeholders, addressing supply chain risks, and ensuring financial transparency.
d. Supporting local charities through employee volunteer programs.

Correct Answer

c. Engaging with stakeholders, addressing supply chain risks, and ensuring financial transparency.

Why?

This choice is the most comprehensive, covering multiple aspects of ESG social responsibility rather than focusing on a single issue.

These educated guessing strategies can significantly improve your chances of selecting the correct answer when you are uncertain about the correct answer. By identifying patterns, avoiding extreme values, and recognizing broad concepts, you can approach multiple-choice exams more strategically.

DECIPHERING AND UNDERSTANDING MULTIPLE-CHOICE EXAM QUESTIONS

Principle 1

The first principle in test taking is to distinctly understand what the question is asking. It is important to have the question's main objective clearly in your mind before going on to consider any of the alternative answers.

Think carefully about what the question is asking and use logic, problem-solving, and critical thinking to eliminate incorrect alternatives.

Checklist

Upon being presented with the question, have you asked yourself ...

- What is this question asking?
- What are you being asked to find?
- What do you need to know to find the answer?
- Is there a specific concept that is being tested?
- What is the concept?
- What information has been provided in either the question's stem or the alternatives that will help me to find the answer?

Principle 2

Operate under the expectation that the question is direct and to the point. Do not assume that the question is designed to trick you. A well-designed, constructed, formatted question will be clear, concise, and to the point... no trickery is intended nor needed. Read the question (and the language used) carefully.

Answer the multiple-choice question solely based on the information given in the question. Base your selection on one of the alternative answers provided solely on the information contained in the question's stem. Do not read into, add to, or embellish the question's stem with spurious or supplemental information.

Look for keywords or phrases to help you better understand the question. Understand exactly what the question stem is asking *before* reviewing or considering the alternative answer choices.

Checklist

Upon reading the question ...

- Have you read the stem correctly?
- Is there a particular way in which the question is phrased?

- Is the question's stem asking for the right (positive) or wrong (negative) response?
- Are you certain that when you read the question's stem, you did not add any spurious content, even subconsciously, to the question's stem?
- Is the question that is being asked clear in your mind?
- Do you have a clear understanding of what the correct answer should be before selecting and recording your answer?

Principle 3

To more accurately assess the question stem and interpret the question correctly, rephrase the question in your own words so that it is very clear in your mind.

Remember Principle 2 ... do not "add" words or alternative meanings into the question's stem.

Checklist

- Have you rephrased the question's stem in your own words?
- Are you certain that by rephrasing the question's stem, you have not added any spurious content, even subconsciously, to the question's stem?
- Is the question that is being asked clear in your mind?
- Do you have a clear understanding of what the correct answer should be?

Principle 4

A fundamental strategy for answering a multiple-choice question correctly is to eliminate the distractors. Eliminating the distractors from the alternative answers listed allows you to then identify the correct answer, which naturally is the alternative that is not a distractor.

A successful approach to identifying the correct answer, hiding among the distractors, is to read the question's stem and each alternative answer *as a pair*, as though the question and the answer are presented as if they were a true or false question.

The following example is a multiple-choice question that demonstrates the fundamental strategy of eliminating distractors by treating each answer choice as if it were a true or false question. This question is based on the "Special Report: ESG Under Strain"[6] by the Thomson Reuters Institute.

Which of the following statements best represents a key challenge discussed in the Special Report: ESG Under Strain?

A. ESG regulations are universally standardized across all countries, eliminating the need for organizations to adjust their practices.
B. Businesses face increasing scrutiny from regulators and stakeholders regarding the authenticity of their ESG commitments.

C. Companies are moving away from ESG reporting due to a complete lack of investor interest in sustainability metrics.
D. The ESG framework primarily focuses on financial performance, with little consideration for environmental or social impacts.

Correct Answer

B. Businesses face increasing scrutiny from regulators and stakeholders regarding the authenticity of their ESG commitments.

How This Demonstrates the Strategy

By treating each option as a true or false statement, we can eliminate the distractors:

A. Incorrect: ESG regulations are not universally standardized; different regions have varying regulations.
B. Correct: The report highlights the growing scrutiny of ESG claims from regulators and stakeholders.
C. Incorrect: Investors still show strong interest in ESG, and reporting continues to evolve.
D. Incorrect: The ESG framework encompasses Environmental, Social, and Governance aspects, not just financial performance.

By eliminating incorrect statements, we identify (b) as the correct answer.

Using the same Reuters document, here is another example of applying Principle 4 to deciphering and correctly answering multiple-choice questions.

What is one of the primary reasons why ESG reporting faces increasing challenges, as discussed in the Special Report: ESG Under Strain?

A. Companies struggle with ESG reporting due to a lack of clear and consistent regulatory frameworks across different jurisdictions.
B. ESG reporting has become completely optional for all publicly traded companies, leading to decreased transparency.
C. The ESG framework is primarily used for marketing purposes and has no real impact on corporate decision-making.
D. Companies have completely abandoned ESG initiatives due to a lack of consumer and investor interest.

How This Demonstrates the Strategy

By treating each answer choice as a true or false statement, we systematically eliminate the distractors.

A. Correct: The report highlights regulatory inconsistencies as a significant challenge, making this the best answer.

B. Incorrect: Many companies are still required to report ESG data, and in some regions, reporting mandates are increasing.
C. Incorrect: While some companies may engage in ESG-washing, ESG frameworks do influence corporate policies beyond just marketing.
D. Incorrect: ESG remains a major focus for businesses due to investor pressure, regulatory developments, and consumer expectations.

This approach ensures that by eliminating false statements, the true statement (a) naturally stands out as correct.

Checklist

- Have you read the question's stem and each alternative answer as individual statements?
- Does the question/alternative answer pair make a false statement? (thereby a distractor)
- Does the question/alternative answer pair make a true statement? (thereby the correct answer)

Principle 5

When attempting to isolate the distractors, among the four alternatives provided, isolate what is important in the alternatives from what is not important relative to the question's stem.

In other words, all alternatives presented may be correct; however, only one is the right choice, the correct answer, for *the specific* question that is being asked. This forces you to keep looking at the question's stem.

Using the document from The Council and European Parliament, "Regulation On Environmental, Social, and Governance (ESG) Rating Activities,"[7] the following multiple-choice question demonstrates the strategy of isolating distractors.

According to the Regulation on Environmental, Social, and Governance (ESG) Rating Activities, what is a primary goal of this regulation?

A. To completely standardize all ESG rating methodologies across the European Union.
B. To ensure transparency and integrity in ESG rating activities while allowing methodological diversity.
C. To require all ESG ratings to focus solely on environmental factors, excluding social and governance aspects.
D. To eliminate the use of ESG ratings by financial institutions due to inconsistencies in their methodologies.

Correct Answer

B. To ensure transparency and integrity in ESG rating activities while allowing methodological diversity.

How This Demonstrates the Strategy

By treating each answer choice as a true or false statement and comparing (isolating) them to the question's stem, you are then able to eliminate distractors:

A. Incorrect: The regulation does not require complete standardization of ESG methodologies. It acknowledges the need for transparency but allows for different approaches.
B. Correct: The regulation focuses on transparency and integrity while still allowing for different ESG methodologies, making this the best answer.
C. Incorrect: ESG ratings must consider Environmental, Social, and Governance factors, not just environmental aspects.
D. Incorrect: The regulation does not seek to eliminate ESG ratings but aims to improve their transparency and reliability.

Applying the Principle 5 approach forces you to carefully isolate what is important (regulatory goals) in this example article from what is not relevant (extreme or incorrect statements), leading to the correct answer.

Checklist

- Have you examined (isolated) each alternative answer as being either true or false concerning the *specific question* that is asked in the question's stem?

Principle 6

After choosing the correct answer and separating it from the distractors, go back to the question's stem and make sure your choice does answer the question.

A successful strategy in identifying the correct answer among the alternative answers presented is to evaluate each alternative answer choice, against the question's stem, independently, not against each other (evaluate the question against each alternative answer). This process will eliminate choosing an answer that does not fit the question.

The World Business Council for Sustainable Development (WBCSD), Executive Vice President Diane Holdorf's article, "Landscape Action: A Business Imperative for Supply Chain Resilience,"[8] serves as the basis for the following example that demonstrates the application of Principle 6.

What is the primary reason businesses should adopt a landscape approach to supply chain resilience, as discussed in the article?

A. It allows companies to work independently and maintain full control over their supply chains.

B. It enables businesses to address multiple sustainability challenges collaboratively beyond their direct supply chains.
C. It eliminates the need for corporate climate targets and government regulations.
D. It focuses solely on financial investments, disregarding environmental and social impacts.

Correct Answer

B. It enables businesses to address multiple sustainability challenges collaboratively beyond their direct supply chains.

How This Demonstrates the Strategy

1. Returning to the Question's Stem
   - The question asks for the *primary* reason businesses should adopt a landscape approach to supply chain resilience.
   - The article emphasizes that landscape approaches allow businesses to collaborate beyond their direct supply chains to address sustainability challenges such as deforestation, land degradation, and climate resilience.
2. Evaluating Each Answer Choice against the Stem
   A. It allows companies to work independently and maintain full control over their supply chains.
      This is *incorrect* because the article stresses collaboration rather than independent action.
   B. It enables businesses to address multiple sustainability challenges collaboratively beyond their direct supply chains.
      This is **correct** because it directly reflects the article's key argument: that landscape approaches foster multi-stakeholder collaboration to solve environmental, social, and economic issues.
   C. It eliminates the need for corporate climate targets and government regulations.
      This is *incorrect* as the article highlights that corporate climate targets (such as those from the Science-Based Targets Initiative) and policy support are essential.
   D. It focuses solely on financial investments, disregarding environmental and social impacts.
      This is *incorrect* because the article emphasizes the integration of financial, environmental, and social factors in supply chain resilience.

By applying the strategy – checking the correctness of each option against the question, rather than comparing them against each other – the best answer (B) becomes clear.

Checklist
Have you ...

- Reread the question's stem.
- Read each alternative answer independently, determining if it answers the question presented.
- Identified an alternative answer that answers the question's stem positively versus negatively.
- Eliminated all alternative answers that provide a false response to the question's stem.
- Selected the alternative answer, which only answers the question's stem positively or completes a true statement.

> Examinations are the only way to know something, at least for a few days.
>
> *unknown*

Principle 7

Identify relationships in two or three of the possible answers (alternatives). Keep in mind that the primary purpose the distractor answers among the possible answers is to "distract" you from identifying and selecting the one correct answer.

When reading the possible answers, if you identify one answer that is considerably different from the others, it is *probably not* the correct choice. Thus assisting you in further eliminating an incorrect response.

The following multiple-choice question, taken from the article, "Regulation On Environmental, Social, and Governance (ESG) Rating Activities,"[9] demonstrates the application of Principle 7.

Which of the following is a key requirement for ESG rating providers to ensure independence and avoid conflicts of interest?

A. They must disclose all potential conflicts of interest to ESMA and establish policies for their prevention and mitigation.
B. They are allowed to offer consulting services to rated entities to improve their ESG scores.
C. ESG rating providers can appoint members of the administrative board of another ESG rating provider within their industry.
D. They may provide investment services along with ESG ratings to offer more comprehensive financial advice.

Correct Answer

A. They must disclose all potential conflicts of interest to ESMA and establish policies for their prevention and mitigation.

How This Demonstrates the Strategy

This question applies the strategy of evaluating each answer independently against the question stem.

- Answer "A" is correct because it aligns with the regulatory requirement for ESG rating providers to disclose conflicts of interest and take measures to prevent them.
- Answer "B" is incorrect because ESG rating providers are explicitly prohibited from providing consulting services to rated entities.
- Answer "C" is incorrect as ESG rating providers are prohibited from exercising significant influence over other rating providers.
- Answer "D" is incorrect because ESG rating providers must not offer investment services to avoid conflicts of interest.

Applying the concepts advanced in Principle 7 ensures that distractor answers are identified and eliminated systematically.

Checklist

Have you:

- Identified a possible answer that is markedly different from all other answers?
- Read the question's stem and each answer as individual statements.
- Examined this answer as being either true or false, to and concerning, the question's stem?
- Selected the answer, which only answers the question's stem positively or completes a true statement.

Principle 8

If two choices begin with the same sound or contain distinctive sounds, spelling, or similarities, the correct answer tends to be one of these two choices.

The example multiple-choice question, using the article "Addressing agricultural scope 3 emissions: Best-practice principles for Within Value Chain Mitigation"[10] illustrates the concept of Principle 8.

What is a key factor in ensuring effective Within Value Chain Mitigation (WVCM) activities according to the document?

A. Farmer facilitation programs – These ensure producers understand and engage with WVCM initiatives.
B. Financial flexibility for farmers – Offering multiple financial structures to farmers ensures long-term adoption.
C. Farmer feedback mechanisms – Establishing trust and transparency through regular communication.

D. Flexible financial formats – Ensuring WVCM project funding adjusts to market changes.

Correct Answer

C. Farmer feedback mechanisms – Establishing trust and transparency through regular communication.

How This Demonstrates the Strategy

1. Two Answers with Similar Sounds or Spelling:
   - "Farmer-led facilitation" (A) and "Farmer feedback mechanisms" (C) both begin with "Farmer," making them likely candidates for the correct answer.
   - "Financial flexibility" (B) and "Flexible funding models" (D) both contain the word "Flexible," making them sound similar and causing potential distraction.
2. Narrowing Down the Correct Answer:
   - While A and C both emphasize farmer involvement, C is correct because it specifically addresses transparency, communication, and trust-building, which are emphasized in the document.
   - B and D focus on financial structures, which are important but secondary to ensuring fairness through engagement.

Checklist

- Can you identify one or more alternative answers that contain distinctive sounds, spelling, or similarities?
- Have you examined each of these alternatives as being either true or false, with respect to and in relation to the question's stem?
- Have you selected the answer, among the two possible alternatives, containing distinctive sounds, spelling, or similarities, which answers the question's stem positively or completes a true statement?

Principle 9

Avoid answers that repeat words contained in the question's stem.

This is often a tricky strategy for test-takers to follow, as many times the multiple-choice question presented will be reworded, specifically in an attempt to examine the test-taker's critical thinking skills to identify the correct answer.

The following example uses the article "State-Owned Enterprises and Private Equity in the Context of Economic Development"[11] to illustrate the application of the important strategy.

Which factor significantly contributes to the increasing presence of State-Owned Enterprises (SOEs) in global markets?

A. Their reliance on state ownership to drive expansion
B. Their ability to align government priorities with financial performance
C. Their dependence on state funding for business growth
D. Their function as state-controlled organizations in economic sectors

Correct Answer

B. Their ability to align government priorities with financial performance.

How This Demonstrates the Strategy

This multiple-choice question demonstrates the strategy of avoiding repeated words from the question stem in the following ways:

1. Avoids Repeating Words from the Question Stem in the Answer Choices
   - The stem contains "**State-Owned Enterprises (SOEs)**" and "**global markets.**"
   - The incorrect choices (**A, C, and D**) **overuse variations of "state"** (**state ownership, state funding, state-controlled**), making them redundant and unhelpful.
   - The correct answer (B) **avoids repeating "state"** and instead uses "**government priorities**" and "**financial performance,**" ensuring it remains distinct.
2. Forces Test-takers to Focus on Meaning Rather Than Wording
   - If a test-taker simply scans for familiar words from the question stem (**state-owned, global markets, etc.**), they might incorrectly choose A, C, or D.
   - Since **B avoids word repetition**, it encourages test-takers to **understand the content rather than relying on pattern recognition.**
3. Why Repetition is a Poor Strategy
   - If answer choices overuse the same words as the question, **test-takers may gravitate toward familiar phrasing rather than engaging with the question's meaning.**
   - This example exposes that trap, making it an effective demonstration of the multiple-choice exam strategy in action.

This question explicitly demonstrates the exam strategy of avoiding repeated words in both the question stem and answer choices.

Checklist

- Have you identified possible answer choices that contain specific or duplicative words used in the question's stem?
- Have you examined each alternative, especially any alternative answer(s) containing repetitive or duplicative words, which may

be found in the question's stem, as being either true or false, with respect to and in relation to the *specific question* that is asked within the question's stem?

- Does the answer that you have selected answer the *specific question* that is asked within the question's stem?

Principle 10

Tell yourself the answer to the question being asked *before* you look at the possible answers that are provided.

Your mind is typically the most focused immediately after you have read the question and digested its contents. At this point, try to predict what the correct answer will be.

Sometimes, the question is too vague for you to anticipate the answer ahead of time. But often, the question's stem is a question in and of itself and may be precise enough for you to answer the question before you even look at the alternative answers.

The following multiple-choice question demonstrates the strategy of predicting the answer before looking at the choices using content from the document. The document used to develop this example question is "ESG and Sustainability Insights: 10 Things That Should Be Top of Mind in 2025."[12]

What challenge is expected to arise as companies work toward net-zero commitments in 2025?

A. Increasing competition from fossil fuel industries
B. The complexity of accurately measuring and reducing Scope 3 emissions
C. Government bans on net-zero initiatives in key markets
D. A decline in investor interest in sustainability efforts

Correct Answer

B. The complexity of accurately measuring and reducing Scope 3 emissions

How This Demonstrates the Strategy

1. Encourages the test-taker to predict the answer before looking at the choices
   - The **question stem is clear and specific** enough that someone familiar with ESG challenges should immediately think about **measurement difficulties, particularly Scope 3 emissions.**
   - Before even looking at the answer choices, a well-prepared test-taker should recall that Scope 3 emissions (indirect greenhouse gas emissions that occur throughout a company's value chain, both upstream (e.g., supply chain activities) and

downstream (e.g., product use and disposal), excluding emissions from owned operations or purchased energy) are the **hardest to track and reduce.**

2. Prevents misleading choices from swaying the test-taker
   - Without predicting first, a test-taker might **hesitate between A, C, or D,** as they sound plausible.
   - However, if they have already thought about **Scope 3 challenges,** they will recognize **B as the strongest choice.**
3. Directly aligns with the document's content.
   - The **document explicitly discusses the challenges of Scope 3 emissions in net-zero strategies.**
   - The question and answer reflect real-world concerns rather than being distracted with **unlikely or misleading options.**

Effectiveness As An Example of the Principle

The example multiple-choice question effectively demonstrates Principle 10 as it:

- Follows the test-taking strategy by encouraging the test-taker to think of an answer before reading the options.
- Minimizes distractions by ensuring the correct answer is logically predictable based on real ESG challenges.
- Uses clear, concise wording that aligns with the document's explanation of the issue.

Checklist

- What is the specific concept that this question is testing?
- Do you have a clear idea of what this question is asking?
- Are you certain that when you read the question's stem, you did not add any spurious content, even subconsciously, to the question's stem?
- Have you composed what the answer should be based on what the question is asking before you looked at the answers?

Principle 11

Identify and address qualifiers in the answer choices provided. Qualifiers are words that alter a statement. Words like always, most, equal, good, and bad.

In multiple-choice questions, **qualifier families** refer to groups of words or phrases that modify a statement by indicating degree, frequency, exclusivity, or possibility. These qualifiers influence whether an answer choice is correct or incorrect by making a statement overly broad, overly restrictive, or appropriately nuanced.

Typical Qualifier Families in Multiple-Choice Questions

Test designers employ qualifier families in designing multiple questions. They assist in crafting well-balanced multiple-choice questions that challenge critical thinking and avoid misleading absolutes. The test taker should be familiar with the following common qualifier families, listed below and categorized based on their function.

1. Absolute Qualifiers (Extreme or Universal Statements)
   These words suggest no exceptions and often make a statement incorrect unless the topic truly allows for no exceptions.
   Examples:
   - Always
   - Never
   - Every
   - All
   - None
   - Must
   - No
2. Frequency Qualifiers (Relative or Time-Based)
   These indicate how often something happens, making statements more or less likely to be true.
   Examples:
   - Usually
   - Frequently
   - Often
   - Sometimes
   - Rarely
   - Occasionally
3. Degree Qualifiers (Intensity or Magnitude)
   These modify the strength of a statement.
   Examples:
   - Completely
   - Entirely
   - Significantly
   - Somewhat
   - Partially
   - Minimally
4. Probability Qualifiers (Possibility or Uncertainty)
   These indicate the likelihood of something occurring, affecting how definitive a statement is.
   Examples:
   - Probably
   - Likely

- Possibly
- May
- Could

5. Exception Qualifiers (Exclusive or Limiting)
   These introduce exceptions that can make an otherwise correct answer incorrect.
   Examples:
   - Except
   - Unless
   - Only
   - Solely
6. Comparison Qualifiers (Relative Differences)
   These words indicate comparison or relativity, which affects the accuracy of a statement.
   Examples:
   - More
   - Less
   - Greater
   - Better
   - Worse

In a multiple-choice test, qualifiers can make an option on a test question be a correct option or an incorrect option.

Using the article "Advancing the S in ESG: A Primer for CFOs,"[13] the following multiple-choice example illustrates the strategy highlighted in Principle 11.

Which of the following best describes the role of CFOs in integrating social performance into business strategy, as discussed in the article?

A. CFOs always prioritize financial performance over social considerations.
B. CFOs never need to engage with investors or regulators on social performance.
C. CFOs could play a key role in ensuring ESG considerations are integrated into enterprise risk management and reporting.
D. CFOs are solely responsible for implementing social performance initiatives without input from other stakeholders.

Correct Answer

C. CFOs could play a key role in ensuring ESG considerations are integrated into enterprise risk management and reporting.

How This Demonstrates the Strategy

The strategy behind Principle 11 is substantiated by:

- Use of Qualifiers: The question includes qualifiers like always, never, could, and solely to highlight how they change the meaning of a statement.
- Impact on Correctness: Answers with absolute qualifiers (always, never, solely) are incorrect because they overgeneralize CFO responsibilities.
- Encouraging Critical Thinking: By presenting different levels of certainty, the question forces test-takers to carefully consider nuanced interpretations.
- Reference to Source Material: The correct answer aligns with the referenced document, reinforcing comprehension.

For further insight on qualifiers and their impact on potential answer choices to a multiple-choice question, see Table 2.1.

Table 2.1 Qualifier Impact on Potential Answer Choices to Multiple-Choice Question

Qualifier	*Importance*	*Effect on Answer Choice*
All	Indicates an absolute condition; be cautious as absolutes are often incorrect. Commonly used but can be easily eliminated if one option is identified as true or false.	Often incorrect. Encourages test-takers to focus on eliminating options rather than understanding the content.
Always	Denotes an absolute condition; often incorrect due to exceptions.	Often incorrect. Makes the distractor less plausible since few things are universally true or false.
Bad	Indicates a negative outcome; can mislead by suggesting an extreme case.	May steer the test-taker toward a negative response even if it's not entirely correct.
Best	Suggests the most effective or appropriate choice. Indicates a superior choice; often used to make one option seem stronger	Can lead test-takers to prefer an answer that sounds more appealing.
Cause	Denotes causation; ensure that one thing directly causes another. Suggests a causal relationship; can mislead if correlation is mistaken for causation.	May cause confusion if causation is incorrectly assumed.
Compare	Requires identifying both similarities and differences between concepts.	Distractors may omit one aspect (e.g., only similarities or only differences), making them less accurate.
Contrast	Focuses solely on differences between concepts.	Distractors may include similarities, which can mislead test-takers who don't read carefully.

(Continued)

Table 2.1 (Continued) Qualifier Impact on Potential Answer Choices to Multiple-Choice Question

Qualifier	*Importance*	*Effect on Answer Choice*
Could	Possibility indicator; not definitive.	Indicates possibility, making it vague but plausible.
Decrease	Indicates a reduction or decline; may be used to mislead if the question is about change rather than direction.	May cause confusion if the actual answer refers to an increase or a different form of change.
Define	Calls for precise meanings of terms.	Distractors often include vague or overly broad definitions to confuse test-takers.
Directly	Refers to a direct cause-and-effect relationship. Implies a strong, immediate connection; may be used to exclude indirect relationships	Can mislead by making indirect relationships seem incorrect.
Effect	Implies an effect; ensure it logically follows the cause. Indicates a result or consequence; may be misleading if multiple effects exist.	May oversimplify complex cause-and-effect relationships.
Enumerate/ list	Involves listing key points or ideas.	Distractors may include irrelevant items, which can be eliminated by careful reading.
Ensures	Strong assertion; often incorrect.	Overly strong statement; makes it likely incorrect.
Equal	Suggests balance or fairness; may be used to mislead if the question requires a distinction.	Can obscure differences and lead to incorrect generalizations.
Evaluate	Requires judgment supported by evidence.	Distractors may lack supporting evidence or present biased judgments, reducing plausibility.
Except	Signals an exception; look for the one option that does not fit. Indicates an exclusion; changes the question's intent.	Reverses meaning; requires careful reading.
False	Indicates inaccuracy; just one incorrect element can make a statement false. Negates a statement; often used in trick questions requiring careful reading	Often incorrect. Requires careful reading; changes the meaning of a question
Good	Indicates a positive outcome; often used to create a false sense of correctness.	May make an incorrect answer seem more appealing than it actually is.
Guarantee	Strong assertion; often incorrect.	Overly strong statement; makes it likely incorrect.

(Continued)

Table 2.1 (Continued) Qualifier Impact on Potential Answer Choices to Multiple-Choice Question

Qualifier	*Importance*	*Effect on Answer Choice*
Illustrate	Demands explanation through examples or analogies.	Distractors might provide examples unrelated to the concept, misleading those with partial knowledge.
Increase	Indicates a rise or growth; may be used to mislead if the question is about change rather than direction.	May cause confusion if the actual answer refers to a decrease or another form of change.
Indirectly	Suggests an indirect relationship; can make an answer seem more complex or uncertain.	May lead to misinterpretation by introducing ambiguity in the relationship.
Justify	Requires reasoning or evidence for a statement.	Distractors often omit logical reasoning, making them easier to identify as incorrect.
Least	Refers to the smallest amount; ensure the answer is the least applicable. Indicates minimal influence or importance; often used to divert attention.	Can make an answer seem unimportant when it may still be relevant.
Less Likely	Suggests probability; often correct.	Suggests a lower probability; often valid in context.
May	Possibility indicator; makes the answer plausible.	Introduces uncertainty; can be correct.
Most	Implies a majority; consider if the answer applies to most cases. Indicates a dominant factor; often used to make a choice seem more significant.	Often correct. May steer test-takers toward an answer that appears more significant.
Most likely	Suggests probability; often correct.	Suggests a high probability; often leads to the right answer.
Never	Indicates an absolute negation; be wary as few things are never true.	Often incorrect. Makes the distractor less plausible since few things are universally true or false.
None	Suggests that no exceptions exist; consider if the statement is too extreme. Absolute term; suggests total exclusion; often incorrect in real-world contexts.	Often incorrect. Can be misleading if exceptions exist.
None of the above	Commonly used but can be easily eliminated if one option is identified as true or false.	Encourages test-takers to focus on eliminating options rather than understanding the content.
Not	Indicates negation; make sure to focus on what is being excluded.	Requires cautious reading; negates meaning.
Often	Relative term; frequently correct.	Suggests frequency; often valid.

(*Continued*)

Table 2.1 (Continued) Qualifier Impact on Potential Answer Choices to Multiple-Choice Question

Qualifier	*Importance*	*Effect on Answer Choice*
Only	Restrictive term; often misleading.	Excludes possibilities, making it misleading.
Primarily	Implies the primary reason or purpose; consider what is most significant. Suggests primary importance; may mislead by implying exclusivity.	May exclude other valid possibilities, leading to an incorrect choice.
Prove	Demands logical or mathematical demonstration of truth.	Distractors may lack logical coherence, which can help identify them as incorrect.
Should	Directive term; suggests advice rather than fact.	Gives advice rather than stating fact.
Some	Refers to a portion of the whole; usually less extreme than "all" or "none." Indicates partial inclusion; often used to make a distractor seem plausible.	Allows flexibility in interpretation, making it more likely to be correct than absolute terms.
Summarize	Calls for a concise account of main ideas.	Distractors might include excessive detail, making them less concise and easier to spot as incorrect.
True	Confirms correctness; ensure all components of the statement are accurate. Affirms a statement; may be used to test knowledge of factual accuracy.	Often Correct. Encourages careful evaluation of factual statements.
Usually	Relative term; frequently correct.	Indicates general trends; usually correct.
Worst	Denotes the least effective or least appropriate choice. Indicates an inferior choice; can be misleading by exaggerating a negative aspect.	Often incorrect. Can mislead by making an option seem worse than it actually is.

EXAM PREPARATION STRATEGIES

Preparing for a professional certification exam requires strategic planning, technical readiness, and strong content mastery. Below is a list of effective preparation strategies

1. Understand the Exam Structure and Requirements
   - Strategy: Review the official exam guide, syllabus, format, duration, and passing criteria.
   - Reasoning: Knowing the structure helps in planning study time effectively and prevents surprises on exam day.

2. Create a Study Plan with a Timeline
    - Strategy: Break down study sessions into manageable sections, focusing on weak areas and key topics.
    - Reasoning: A structured approach ensures all topics are covered and prevents last-minute cramming.
3. Utilize Official Study Materials and Practice Exams
    - Strategy: Use study guides, recommended books, and official practice tests to familiarize yourself with question types.
    - Reasoning: Official materials align closely with exam content, increasing familiarity and confidence.
4. Take Online Mock Exams
    - Strategy: Simulate exam conditions by taking timed, full-length practice tests online.
    - Reasoning: Helps with time management, reduces anxiety, and identifies weak areas.
5. Strengthen Technical Readiness
    - Strategy: Test your computer, Internet speed, browser compatibility, and exam software well in advance.
    - Reasoning: Prevents technical disruptions that could impact your performance.
6. Master Time Management
    - Strategy: Allocate time per question based on total exam duration and practice answering within the set time limit.
    - Reasoning: Ensures you complete the test without rushing through critical sections.
7. Review Key Concepts Using Active Recall and Spaced Repetition
    - Strategy: Use flashcards, mind maps, and spaced repetition tools (e.g., Anki) for long-term retention.
    - Reasoning: Active recall improves memory retention and helps with recall under exam pressure.
8. Set Up a Distraction-Free Test Environment
    - Strategy: Find a quiet room, use noise-canceling headphones if needed, and inform others to avoid interruptions.
    - Reasoning: Online proctored exams have strict requirements, and distractions can disrupt focus.
9. Ensure Proper Identification and Compliance with Exam Rules
    - Strategy: Have a valid ID, follow webcam/microphone rules, and review the proctoring guidelines.
    - Reasoning: Not adhering to rules may lead to disqualification or technical issues.
10. Get Enough Rest and Stay Hydrated Before the Exam
    - Strategy: Sleep well the night before and stay hydrated for peak cognitive function.
    - Reasoning: Fatigue and dehydration can negatively impact concentration and performance.

11. Develop Stress Management Techniques
    - Strategy: Practice deep breathing, mindfulness, or visualization exercises to reduce anxiety.
    - Reasoning: Staying calm helps maintain focus and clarity throughout the exam.
12. Join Online Study Groups and Discussion Forums
    - Strategy: Engage with professionals preparing for the same certification in online communities.
    - Reasoning: Group discussions provide insights, clarify doubts, and enhance learning.
13. Read and Understand Exam Policies on Retakes
    - Strategy: Be aware of retake policies, waiting periods, and associated costs.
    - Reasoning: Helps in planning alternative attempts in case the first attempt is unsuccessful.

Proven Strategies to Successfully Answering Multiple-Choice Questions

1. Understand the Question Before Looking at the Options
    - Read the entire question carefully before looking at the choices.
    - Read the question thoroughly: Don't rush. Ensure you fully understand what the question is asking.
    - Identify key terms and watch for tricky (qualifier) words like "not," "except," and "always."
    - Identify keywords in the question to focus your attention.
    - Answer in your head first: Try to formulate the answer in your mind before looking at the options to avoid being influenced by incorrect answers.
    - Break long questions into smaller parts to ensure full comprehension.
    - Watch the meaning of sentences containing double negatives: Eliminate both negatives and then answer the question.
2. Eliminate Incorrect Answer Choices
    - Eliminate answers that are obviously incorrect, illogical, or contradict the information provided to narrow down choices.
    - Be cautious of absolute qualifiers like "always," "never," "all," "none," and "every" – these are usually incorrect.
    - Look for opposites: Sometimes, one of the correct answers will be the direct opposite of another answer choice. Examine opposites carefully: one is definitely incorrect, and one is likely to be correct.
    - Compare answer choices to the stem – grammatical consistency can help rule out incorrect options.
    - Consider the "most inclusive" answer: If multiple options seem close, choose the one that encompasses the most information or covers the widest range of possibilities. These are called "umbrella" options.

- Consider True/False Labeling: Label alternative answers as true or false to find patterns and eliminate incorrect options. If any part of an answer choice is demonstrably false, then the entire answer is incorrect.

3. Answer the Easy Questions First
   - If the testing platform allows you to scan ahead and view all of the test questions, quickly scan the test and answer questions you can readily answer to build confidence and save time.
   - If the testing platform provides the capability, identify difficult questions and return to them later. Skipping and returning to questions prevents getting stuck on one problem.
   - Recycle through the Test: Revisit skipped questions, as answers may come to you more easily after answering other questions or from clues found elsewhere in the exam
4. Guessing Strategies
   - If there is no point (grade) penalty for guessing, always *answer every* question.
   - If two answers seem correct, pick the *more inclusive* one.
   - Choose answers with relative qualifiers like "sometimes," "often," and "usually."
   - When absolutely guessing, select the middle responses (e.g., B or C).
   - If you must guess, look for grammar agreement between the question stem and the alternative answers, and eliminate options that are outside the question's topic.
5. Managing Time Effectively
   - If the testing platform allows, review the exam: Get an overview of the exam to identify easier questions and plan your time.
   - Save time at the end to review any marked/skipped questions.
   - Answer all questions within the given time, and don't spend too much time on any one question.
   - Don't spend too much time on a single question.
   - If you're unsure about a question, move on and come back to it later.
   - Set time goals and pace yourself to ensure you have time to attempt all questions.
6. General Strategic Tips
   - Scientific-sounding options tend to be more correct than casual wording.
   - Longer answers often contain more details and are likely correct.
   - Look for grammatical matches between the question and the answer.
   - Be alert to terminology that links alternatives or questions to key exam content areas.
   - Remember that you are looking for the best answer, not just a correct one.

- Change Answers Carefully: Be prepared to change your answer if you have a **clear** reason but avoid changing based on nervous feelings.
- Even if guessing, make sure to answer every question, especially if there is no penalty for wrong answers.
- Don't read anything in or out of the question: Make sure you restrict your response only to the topic at hand.

SUMMARY

Multiple-choice exams test your critical thinking, problem-solving skills, and judgment.

The key to excelling in multiple-choice exams is understanding the question, eliminating wrong answers, managing time, and making educated guesses when necessary.

Adhering to the principles and strategies provided in this chapter will significantly improve your accuracy and confidence during tests.

We wish you success in passing your certification examination.

The night before an exam is like the night before Christmas. You can't sleep and hope for a miracle too!

unknown

§

NOTES

1 Posey, B., October 19, 2012, "Improving your Odds of Answering Multiple Choice Questions Correctly," www.pearsonitcertification.com/articles/article.aspx?p=1960217, retrieved February 5, 2025.
2 A Selected-Response Assessment exam refers to a testing format where test takers select an answer from given options rather than constructing their own response. A Multiple-Choice Test (MCT) is a common informal term, but "selected-response" is the broader educational term that includes true/false, matching, and multiple-choice items.
3 (n.a.), March 7, 2023, "Advancing the S in ESG: A Primer for CFOs," World Business Council for Sustainable Development (WBCSD), www.wbcsd.org/resources/advancing-the-s-in-esg.
4 (n.a.), March 7, 2023, "Advancing the S in ESG: A Primer for CFOs," World Business Council for Sustainable Development (WBCSD), www.wbcsd.org/resources/advancing-the-s-in-esg.
5 Thomson Reuters, 2022, "Special Report: ESG under Strain," Thomson Reuters Institute, www.thomsonreuters.com/en/reports/esg-under-strain?utm_source=chatgpt.com.

6 The Council and European Parliament, June 13, 2023, "Regulation on Environmental, Social, and Governance (ESG) Rating Activities," https://data.consilium.europa.eu/doc/document/PE-43–2024-INIT/en/pdf.

7 Holdorf, D., January 27, 2025, "Landscape Action: A Business Imperative for Supply Chain Resilience," World Business Council for Sustainable Development (WBCSD), https://tinyurl.com/Holdorf-D-Author.

8 The Council and European Parliament, June 13, 2023, "Regulation on Environmental, Social, and Governance (ESG) Rating Activities," https://data.consilium.europa.eu/doc/document/PE-43–2024-INIT/en/pdf.

9 3Keel, January 2025, "Addressing Agricultural Scope 3 Emissions: Best-Practice Principles for Within Value Chain Mitigation," Department for Energy Security and Net Zero (DESNZ) and Department for Environment, Food, and Rural Affairs (Defra), https://tinyurl.com/3Keel-Article.

10 Jiang, I., Wang, Z., Krosinsky C., and Huang, J., November 22, 2024, State-Owned Enterprises and Private Equity in the Context of Economic Development, *Sustainability Leadership Series,* 2(3), (https://sfini.org), https://img1.wsimg.com/blobby/go/a1525465-9048-4fda-9d12-641522978982/20241122-SFI%20Case%20Studies%20-%20State-Owne-1d3329f.pdf.

11 Davies, P. and Huber, B., January 18, 2025, "ESG and Sustainability Insights: 10 Things That Should Be Top of Mind in 2025," Latham & Watkins LLP, Harvard Law School Forum on Corporate Governance, https://corpgov.law.harvard.edu/2025/01/18/esg-and-sustainability-insights-10-things-that-should-be-top-of-mind-in-2025/.

12 (n.a.), March 7, 2023, "Advancing the S in ESG: A Primer for CFOs," World Business Council for Sustainable Development (WBCSD), www.wbcsd.org/resources/advancing-the-s-in-esg.

13 Figure 2.1, Elements of a Multiple-Choice Question, created by authors, based upon content found in (n.a.), March 7, 2023, "Advancing the S in ESG: A Primer for CFOs," World Business Council for Sustainable Development (WBCSD), www.wbcsd.org/resources/advancing-the-s-in-esg, February 5, 2025.

REFERENCES

This chapter was written and organized, utilizing but not limited to the following additional sources. Custom modification to content adapted from these sources was done to present material more closely and accurately representative of the examination for which the test candidate is preparing to take.

(n.a.), n.d., "Multiple Choice and Other Objective Tests," Butte College, www.butte.edu/departments/cas/tipsheets/studystrategies/multiple_choice.html.

(n.a.), n.d., "Clue Words in Exam Questions," Academic Resource Center, George Fox University, www.georgefox.edu/offices/arc/resources/exam-clue-words.pdf.

(n.a.), n.d., "How to Tackle Exam Questions," Cornell University, The Learning Strategies Center, https://lsc.cornell.edu/how-to-study/studying-for-and-taking-exams/exam-strategies-how-to-tackle-exam-questions-3.

(n.a.), n.d., "Exam Tips," The University of Queensland Australia, https://my.uq.edu.au/information-and-services/student-support/study-skills-and-learning-advice/study-skills-and-learning-advice-overview/exam-tips.

(n.a.), n.d., "How to Take Multiple Choice Tests," Minot State University, www.minotstateu.edu/online/pages/tour_and_skills/skills_05.shtml.

(n.a.), n.d., "Multiple Choice Test Strategies," Academic Success Center, Oregon State University, https://success.oregonstate.edu/learning-corner/test-taking/multi-choice.

(n.a.), n.d., "Five Tips to Ace Your Multiple Choice Exams," Robert Gillespie Academic Skills Centre, University of Toronto Mississauga, www.utm.utoronto.ca/rgasc/media/2073/download?inline.

(n.a.), n.d., "Strategies for Answering Multiple Choice Questions," The University of New Orleans, www.uno.edu/media/10076/download#:~:text=%2D%20If%20you%20aren't%20sure,alternative%20that%20is%20most%20inclusive.

(n.a.), n.d., "Test Taking Strategies," California State University, Fullerton, www.fullerton.edu/aas/_resources/pdfs/Test%20Taking%20Strategies.pdf.

(n.a.), n.d., "Test-Taking Strategies – Common Questions," University of Northern Colorado, www.unco.edu/student-academic-success/academic-success/academic-success-resources/pdf/SAS_test_taking.pdf.

(n.a.), n.d., "Tips for Multiple Choice Exams," University New South Wales, Australia, https://student.unsw.edu.au/multiple-choice-exams.

(n.a.), March 24, 2022, "Clue Words in Exams," University New South Wales, Australia, www.student.unsw.edu.au/clue-words.

(n.a.), July 2016, "Tips for Tests," Universiteit Leiden, www.universiteitleiden.nl/binaries/content/assets/science/onderwijs/sterrenkunde/guidetipsfortests.pdf.

(n.a.), November 2, 2014, "The Art of Multiple Choice in the Era of CCLS," The Teaching Factor, https://theteachingfactor.wordpress.com/tag/text-dependent-questions.

(n.a.), 2008, "Picking Distractors for Multiple Choice Questions," Maplesoft, a Division of Waterloo Maple Inc., www.maplesoft.com/view.aspx?SF=5732%2FUsingDistractors.pdf.

Clay, B., 2001, "Is This a Trick Question? A Short Guide to Writing Effective Test Questions," Kansas Curriculum Center, www.k-state.edu/ksde/alp/resources/Handout-Module6.pdf.

Collins, B., March 13, 2003, "Test-Taking Tactics, Part I, Evaluating Answer Choices," https://m.delphiforums.com/pmfd/messages/298/1.

Keeley, M., 1997, "Test-Taking Strategies," Special Populations Office, Bucks County Community College, https://faculty.bucks.edu/specpop/tests.htm.

Paul, W. and Owens, R., January 9, 2013, *How to Study in College*, 11th edition. Cengage Learning, www.cengage.com.

Posey, B., October 19, 2012, "Improving your Odds of Answering Multiple Choice Questions Correctly," www.pearsonitcertification.com/articles/article.aspx?p=1960217.

Raphael, T., February 1986, Teaching Question Answer Relationships, Revisited. *The Reading Teacher*, 39(6), 516–522, https://lattinr.weebly.com/uploads/4/9/5/3/49530073/teaching_question_answer_relationships.pdf.

Shain, D.D., June 29, 1995, Multiple Choice Test-Taking Strategies. In: *Study Skills and Test-Taking Strategies for Medical Students*. Springer-Verlag Publishers, www.dmu.edu/wp-content/uploads/MULTIPLE-CHOICE-TEST-TAKING-STRATEGIES.pdf.

Shin, J., Guo, Q., and Gierl, M., April 24, 2019, Multiple-Choice Item Distractor Development Using Topic Modeling Approaches. *Frontiers in Psychology*, 10, 825. http://dx.doi.org/10.3389/fpsyg.2019.00825, www.frontiersin.org/journals/psychology/articles/10.3389/fpsyg.2019.00825/full.

The Princeton Review, May 7, 2024, "Princeton Review Digital SAT Premium Prep," 2025, The Princeton Review, www.princetonreview.com/?exid=e61445e5-6884-4529-b722-625acbc5595c&exdt=2&gad_source=1&gclid=EAIaIQobChMIqqnLg_-5iwMVUUT_AR39lQOOEAAYASAAEgJ5IPD_BwE.

Thompson, N., October 17, 2022, "Distractor Analysis for Test Items," Assessment Systems Corporation, https://assess.com/distractor-analysis-test-items.

Van Blerkom, D.L., January 6, 2011, Multiple Choice Test Strategies, In: *College Study Skills: Becoming a Strategic Learner*, 7th edition. Wadsworth Publishing, University of Hawaii Maui College, https://maui.hawaii.edu/wp-content/uploads/sites/27/2015/08/Multiple_Choice_Test_Strategies.pdf.

Vera, M., September 27, 2014, "10 Effective NCLEX Test Taking Strategies and Tips," https://nurseslabs.com/10-effective-nclex-test-taking-strategies-tips.

Chapter 3

Reading for Comprehension and Understanding

Principles and Strategies

> Once thought of as the natural result of decoding plus oral language, comprehension is now viewed as a much more complex process involving knowledge, experience, thinking, and teaching. Comprehension inherently involves inferential and evaluative thinking.
>
> Fielding and Pearson[1]

INTRODUCTION

The National Assessment of Adult Literacy (NAAL) survey conducted in 2003 evaluated the literacy skills of adults in the United States. The survey examined reading (prose literacy), understanding documents, and using numbers (quantitative literacy). The survey categorized adult literacy into four levels: Below Basic, Basic, Intermediate, and Proficient. Most adults (53%) demonstrated Intermediate literacy skills and could read and understand moderately challenging texts.[2]

According to the 2023 Program for the International Assessment of Adult Competencies (PIAAC) survey, there has been a significant decrease in adult literacy scores across many countries, including the United States. In the United States, literacy scores for adults aged 16–65 declined by 12 points from 2017 to 2023. The survey's disclosure that a growing percentage of adults scored at the lowest levels in literacy, jumping upward from 19% in 2017 to 28% in 2023, is concerning. The survey also revealed that globally, approximately one in five adults is a low performer in literacy.[3]

These results suggest that a significant portion of the adult population may struggle with tasks requiring reading comprehension and understanding of written texts.

This chapter uses the terms article and text when referencing the required reading materials for the ESG Certification exam. The broader concept of text adheres to the definition as articulated in The National Assessment of

DOI: 10.1201/9781032672557-3

Educational Progress (NAEP), Reading Framework for the 2026 National Assessment of Educational Progress report, directed by the U.S. Department of Education.

NAEP provides the following definition of text, which guides the usage of this term in this chapter.

> Texts are generated by authors to communicate to readers. Texts take many forms, drawing on multiple genres and combinations of genres. They relay vastly different content to address many kinds of purposes. They draw on a wide array of modalities (e.g., static print, nonlinear hypertext, images, videos), sometimes combining modalities into multimodal forms (e.g., print with images or links to videos). They may be printed on paper or published in digital forms. They also differ in complexity, a term that usually refers to the density and nuance of texts' ideas and language structures.
>
> Texts are composed according to conventions tied to cultural traditions and social practices. These traditions and practices are developed within and across such disciplines as literature, science, or history. Such conventions include genre traditions favored by disciplines and modalities that are selected because of the ways they communicate certain kinds of ideas. Texts also vary in terms of the people, points of view, and experiences that are or are not represented. This means that texts may be readily understood by readers who find the ideas familiar or compelling but are more challenging for others who have no frame of reference or interest in the topic.[4]

The objective of this chapter is twofold. First, to provide the test taker with insights, tips, and strategies to maximize the test taker's reading for comprehension and understanding of the material contained in the texts identified as required reading for the ESG Certification exam. Second, applying these tips and strategies helps strengthen the test taker's ability to more aptly identify the correct answer on the ESG multiple-choice exam questions.

Before heading straight to the tips and strategies, it is important to discuss the basic definition of reading for comprehension and reading for understanding and then examine the two concepts side-by-side. Doing so will provide valuable insight necessary for the test taker when working to recognize, understand, and comprehend the key concepts from each of the required readings for the ESG Certification exam.

To that end, the first step is to clearly define the concepts of reading for comprehension and reading for understanding and then explore each concept further.

WHAT IS READING FOR COMPREHENSION?

Reading for comprehension is the ability to actively process and grasp the meaning of a text, going beyond simply decoding words. It involves integrating new information with existing knowledge, making connections, and understanding the text's significance. This deeper engagement requires critical thinking and enables readers to draw conclusions and apply what they've learned. Ultimately, effective reading for comprehension leads to better retention and a fuller understanding of the material.

A HEURISTIC FOR THINKING ABOUT READING COMPREHENSION

In 1999, the Office of Educational Research and Improvement of the U.S. Department of Education charged the RAND Reading Study Group (RRSG) with developing a research agenda to address the most pressing issues in literacy. The decision was made to focus this research on reading comprehension. The RRSG defined the term reading comprehension as the process of simultaneously extracting and constructing meaning through interaction and involvement with written language.

The RRSG proceeded to develop a heuristic for thinking about reading comprehension that emphasizes the interaction of the reader, the text, the activity, and the context (see Figure 3.1). This heuristic underscores that reading comprehension is not merely a property of the reader or the text alone but rather emerges from a dynamic interplay among these four elements.

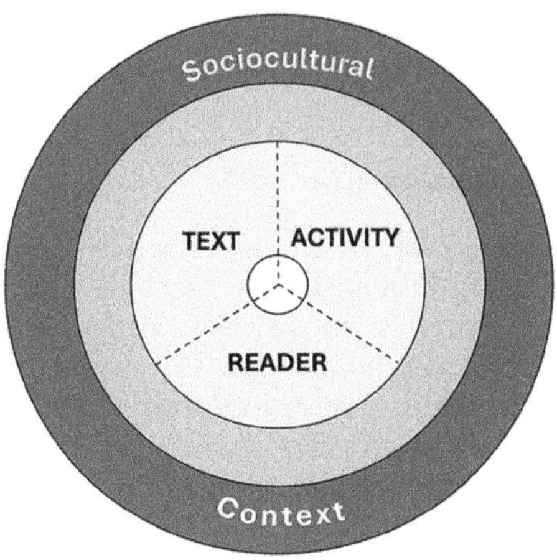

Figure 3.1 A Heuristic for Thinking About Reading Comprehension.[5]

The reader brings their skills, knowledge, and experiences to the text, while the text presents varying levels of complexity, structure, and content. Furthermore, the reading activity itself, including the purpose and goals for reading, and the context in which reading occurs, such as the social, cultural, and instructional environment, all significantly influence the level of understanding achieved.

This model suggests that effective reading comprehension involves a flexible and adaptive process where readers actively construct meaning by integrating information from the text with their own knowledge within a specific activity and context.[6]

WHAT IS READING FOR UNDERSTANDING?

Reading for understanding involves grasping the basic meaning of a text by decoding words and recognizing sentences. It's the initial step in processing information, focusing on the literal meaning and surface-level content. Think of it as summarizing the main points: you know what the text says. While fundamental, reading for understanding may not require deeper analysis or connection to prior knowledge.

While often used interchangeably, "reading for understanding" and "reading for comprehension" have distinct differences in depth of engagement, cognitive processing, and application.

The following material provides the reader with analysis and discussion on these two concepts by examining further the distinct differences between the two concepts.

READING FOR COMPREHENSION VS. READING FOR UNDERSTANDING

Reading for Comprehension

Reading for comprehension is the ability to actively process and grasp the meaning of a text, going beyond simply decoding words. It involves integrating new information with existing knowledge, making connections, and understanding the text's significance. This deeper engagement requires critical thinking and enables readers to draw conclusions and apply what they've learned. Ultimately, effective reading for comprehension leads to better retention and a fuller understanding of the material.

The Components of Comprehension

Comprehension consists of a combination of several components that work together and contribute to an overall understanding of the materials read.

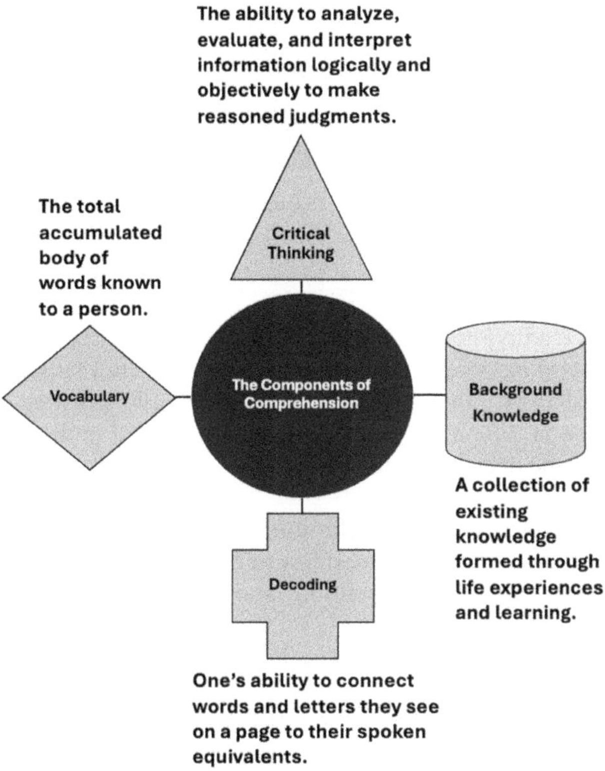

Figure 3.2 The Components of Comprehension.[7]

These components are background knowledge, vocabulary, decoding, and critical thinking. See Figure 3.2.

Summarizing ... reading for comprehension (Basic Grasp)

- Focuses on processing and grasping the surface/basic meaning of the material to be read.
- Involves decoding words, recognizing sentences, and understanding their literal meanings.
- Represents the initial step in processing information.

Example: An individual reads a white paper about climate change and can summarize its main points, such as rising global temperatures and melting ice caps.

Gough and Tunmer, researchers most well-known for developing the "Simple View of Reading," state that reading equals the product of decoding (D) and comprehension. Stated simply as $R = D \times C$, where each variable ranges from 0 (nullity) to 1 (perfection).[8]

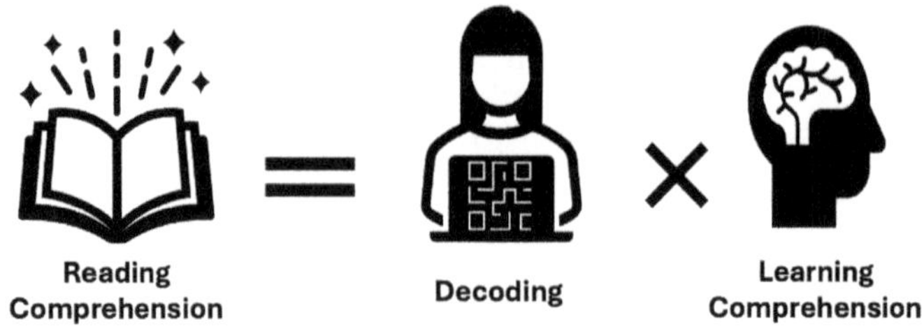

Figure 3.3 Simple View of Reading.[9]

Therefore, reading comprehension, stated as RC, is the product of two key skills: decoding (word recognition) stated as D, and language comprehension stated as LC, giving $RC = D \times LC$, essentially meaning that to read well, one must be able to decode words on a page and understand the language used in the text. See Figure 3.3.

> Real reading has to do with thinking, learning, and expanding a reader's knowledge and horizons. It has to do with building on past knowledge, mastering new information, and connecting with the minds of those you've never met.[10]

Scarborough's Reading Rope

Scarborough's Reading Rope is a comprehensive model that illustrates the complex skills required for proficient reading. Developed by Dr. Hollis Scarborough, it visualizes reading as a rope composed of multiple intertwined strands.

The Reading Rope consists of two main sections:

Word Recognition:

- Phonological awareness
- Decoding
- Sight recognition of familiar words

Language Comprehension:

- Background knowledge
- Vocabulary
- Language structures
- Verbal reasoning
- Literacy knowledge

These strands work together to produce skilled reading comprehension.

The Reading Rope is highly relevant to reading, understanding, and comprehension because it:

- Demonstrates the interconnectedness of various reading skills.
- Highlights that weaknesses in any strand can affect overall reading proficiency.
- Emphasizes the importance of both decoding and language comprehension in achieving reading success.

While Scarborough's Reading Rope is often used to explain reading development in children, it is equally relevant for adult learners. Many adults preparing for professional certifications, such as ESG or finance exams, need to strengthen their comprehension strategies. Even proficient readers benefit from refining skills like verbal reasoning, background knowledge activation, and vocabulary development to engage with complex, technical materials.

To apply the tenets of Scarborough's Reading Rope when preparing for a professional certification exam:

1. Assess your strengths and weaknesses across all strands of the rope.
2. Create mental images or diagrams to visualize complex concepts, strengthening the connection between word recognition and comprehension.
3. Develop verbal reasoning. Practice critical thinking and inference skills through case studies and application-based questions.
4. Enhance background knowledge related to the exam topics. Actively research and build familiarity with key topics related to the exam before diving into dense study materials.
5. Expand your vocabulary. Identify and master industry-specific terminology to improve comprehension.
6. Improve fluency. Read and summarize material efficiently to retain key concepts and make connections across topics.
7. Practice verbal reasoning by explaining concepts aloud or discussing them with peers.
8. Strengthen language structures. Study the format and style of exam questions. Read well-structured, technical writing to develop an understanding of how complex ideas are conveyed.
9. Use metacognitive strategies, such as self-questioning and reflection, to deepen your understanding of the material.

By integrating these elements, professionals can improve their ability to decode, analyze, and retain information, leading to more effective exam preparation and performance.

Reading for Understanding

Reading for understanding involves grasping the basic meaning of a text by decoding words and recognizing sentences. It's the initial step in processing information, focusing on the literal meaning and surface-level content. Think of it as summarizing the main points: you know what the material you have read says. While fundamental, reading for understanding may not require deeper analysis or connection to prior knowledge.

Summarizing ... reading for understanding (Deeper Engagement):

- Goes beyond the surface to analyze, interpret, and connect ideas within the material to prior knowledge.
- Involves higher-order thinking skills, such as critical thinking, evaluation, and drawing inferences.
- Leads to better retention and application of information.

Example: A scientist reads the same article on climate change and evaluates the credibility of the data, compares it with previous research, and considers its implications for policy changes. This involves a deeper engagement, interpretation, and critical thinking beyond what is explicitly stated in the article.

Table 3.1 provides a summary of the key differences between reading for comprehension and reading for understanding.

EXPOSITORY VS. NARRATIVE WRITING

When reading for both comprehension and understanding, it is important to identify the writing "type" of the material to be read.

Table 3.1 Key Differences Summarized: Reading for Comprehension vs. Reading for Understanding

Feature	*Reading for Comprehension (Basic)*	*Reading for Understanding (Deeper)*
Depth	Basic grasp of written material's meaning	Deeper engagement, interpretation, and critical thinking.
Process	Decoding and recognizing	Analyzing, evaluating, and connecting to prior knowledge.
Knowledge	Knowing literal meanings	Integrating information and making connections.
Retention	Potentially temporary	Leads to better retention and application.

According to the Interactive English Language Centre, simply put, a narrative writing is a type of presentation that tells a story or describes a sequence of events. The purpose of a narrative text is to entertain or inform the reader by presenting a series of events coherently and engagingly.[11]

In contrast, expository writing doesn't focus on telling a story or evoking emotions. It's concerned with presenting factual information or explaining a process. The goal is not to entertain but to enlighten the reader on a specific topic, concept, or process.

Expository text is typically structured with visual cues such as headings and subheadings that provide clear cues as to the structure of the information. The first sentence in a paragraph is also typically a topic sentence that clearly states what the paragraph is about.

The expository text also often uses one of five common text structures as an organizing principle:

- Cause and effect
- Problem and solution
- Compare and contrast
- Description
- Time order (sequence of events, actions, or steps)[12]

Being aware of these structures can help the reader to recognize relationships between ideas and the overall intent of the article.

Both expository and narrative texts play crucial roles in developing reading understanding ND comprehension skills, but they engage different cognitive processes.

- Narrative writing enhances readers' abilities to follow plot sequences, understand character motivations, and make inferences about events and outcomes. These skills contribute to emotional intelligence and social understanding.
- Expository writing focuses on assisting readers in developing skills in identifying main ideas, understanding cause-and-effect relationships, and synthesizing information from multiple sources.

Understanding the differences between the two writing types is important. The test taker, when reading the required material for the Certification exam, will encounter expository writing. It is important to identify the main ideas presented along with synthesizing the information from each article into one's own words. This approach will be beneficial in helping to identify the correct answer out of the distractors presented with each multiple-choice question on the exam.

Table 3.2 provides a comparison between these two important writing types.

Table 3.2 Expository vs. Narrative Writing

Aspect	*Expository*	*Narrative*
Purpose	To inform, explain, or describe.	To entertain or tell a story.
Structure	Organized by topic, often with headings.	Chronological sequence with plot elements.
Content	Facts, data, and explanations.	Characters, events, and dialogue.
Language	Clear, concise, fact-based, and objective.	Descriptive, emotive, and figurative.
Text features	Diagrams, charts, and bullet points.	Descriptive passages and character development.
Tense	Usually present tense.	Often past tense.
Examples	Textbooks, news articles, manuals.	Novels, short stories, fables.
Comprehension focus	Main ideas, facts, and logical connections.	Plot, characters, and themes.
Reading skills required	Identifying main ideas, summarizing, analyzing arguments.	Making inferences, understanding themes, tracking character development.

STRATEGIES FOR A BETTER UNDERSTANDING AND COMPREHENSION OF WRITTEN TEXT

In preparation for the ESG Certification exam, the test taker will read required texts covering diverse topics aligned with the three ESG pillars.

Many of these texts address information that may be fundamental knowledge to some, while others will find the texts completely foreign. Whether your expertise across the pillars of ESG is extensive or ESG is a new venture, everyone will benefit from additional coaching in increasing their understanding and comprehension of the ESG-related material read.

A better understanding and comprehension of what is read enhances subject matter knowledge and retention while heightening the ability to apply that knowledge. Subject understanding, comprehension, and retention are essential in identifying both the distractors and the correct answer in multiple-choice question exams.

Reading is vital for mental development and growth. Reading comprehension can help improve focus and concentration. Research shows that the simple act of understanding what you read can help to improve your attention span. When you comprehend the material, your brain can better filter distractions and focus on the task.

Research shows that the strategies listed below will lead to better problem-solving and critical thinking skills, memory, focus, empathy, reading fluency, and writing strengths.[13]

Based on recent research and expert recommendations, listed below are effective techniques for improving reading, understanding, and comprehension. Test takers are encouraged to review the following techniques to adapt them to fit individual learning styles, reading, and comprehension levels, in preparation for the ESG Certification exam.

Understanding and Comprehension Strategies

1. **Ask Questions**
   While reading, ask yourself questions about the material you have read. What is the main idea? What evidence does the author use? This tactic helps you stay focused and encourages deeper engagement with the material.
2. **Connect with the Topic**
   Develop skills that enable you to turn the text from something that's being read to material that means something to you. Personally, engaging with the text being read can spark empathy and excitement, fostering a deeper interest that enhances both the desire and ability to understand the material being read.
3. **Draw Conclusions**
   Draw conclusions or infer ... make educated guesses based on information provided in the text combined with your prior knowledge and experiences. A reader makes inferences by establishing appropriate, meaningful connections between separate pieces of information stated in the text and the reader's background knowledge. It's often described as "reading between the lines."[14]
4. **Engage in Discussions**
   It's crucial to talk about texts to truly understand them. When you discuss what you've read, you get to voice your thoughts. This is your chance to clarify anything that seems confusing or unclear. You'll also hear how others interpreted the text, which might challenge your views and give you new perspectives you hadn't considered.
   This active way of learning helps solidify the information in your mind while you actively construct knowledge, which strengthens your recall and helps you apply what you've learned more effectively. This also boosts your critical thinking skills, allowing you to recall and use information more effectively. So, don't just read – discuss, question, and engage!
5. **Hone Accurate and Automatic Reading Skills**
   Developing the ability to read words accurately and automatically is essential for fluent reading, which, in turn, supports better comprehension.[15]
6. **Identify Main Idea/Message**
   Making a conscious effort to identify the main idea of a passage can help improve reading comprehension. What is this article about? Why

does it matter? Being able to identify the subject of an article quickly can help you get a clear picture and understanding of the point of what you are reading.

7. **Predict a Direction**

   Predicting encourages readers to go beyond the material they're reading to make assumptions and expectations about what is coming next. In the context of reading comprehension, readers need to think critically and problem-solve using building blocks such as background knowledge, vocabulary, and text structure to decide where they think the article is heading. This will keep you engaged in the material and encourage you to look beyond the surface-level content. Doing so will allow you to gain a deeper understanding and comprehension of the material.

8. **Preview the Material**

   Before diving deep into an article, spend a few moments previewing the content. Look at the headings, subheadings, and any highlighted or bolded words. Skimming the content prepares your mind for the material it's about to encounter and activates prior knowledge.

9. **Summarize, Paraphrase, and Synthesize**

   Summarizing focuses on restating key points. Regularly pause to summarize what you've read, *in your own words*. Paraphrasing helps to reinforce understanding and ensures that you grasp the article's main points. When you use your own words to describe what you have read, it makes it more relatable and more memorable to you. Synthesizing refers to the process of combining new information from a text with prior knowledge to create a deeper, more complete understanding of the material. It involves reflecting on evolving ideas throughout the reading process, identifying connections between different parts of the text (or multiple texts), and forming original insights, perspectives, or conclusions. Synthesizing what you have read into "knowledge bites" will require you to actively integrate information and think critically to develop a new "whole" or understanding of the material read.

10. **Think about Thinking**

    The scientific term is metacognition, the process of thinking about how we think and learn. It is essentially exercise for the brain. In addition to helping the reader make connections to self, other text, and the world, metacognition allows for the reader to plan, monitor, and activate strategies when new information is presented. The engagement between our thinking, reflecting, and verbalizing leads to a deeper understanding of the text. Through the think-aloud process, the reader examines the mental images that are directly connected to life experiences, background knowledge, and memories.

    A specific strategy that a test taker may use to better prepare for an exam, based on the concept of metacognition, is **self-questioning**

and reflection. This involves actively thinking about what they know, identifying gaps in their knowledge, and using strategies to strengthen their understanding.

Application:

While reading about ESG reporting standards (e.g., GRI or SASB), for example, the professional verbalizes their thought process:

This reminds me of the sustainability report I reviewed at work last quarter. The company used GRI standards to disclose its carbon emissions. How does this align with the materiality principle in ESG reporting?

By connecting new information to prior knowledge or professional experiences, they deepen their understanding of the material.

11. **Utilize Background/Prior Knowledge**

    Connecting and drawing upon information, experiences, and understanding that you may have, e.g., background or prior knowledge (even if limited) of the subject matter, even before you read an article, will assist you in achieving a better and sustained comprehension of the material.

12. **Visualizing as You Read**

    Visualizing in reading comprehension refers to the process of creating mental images based on the text being read. This strategy involves using sensory details from the text to imagine scenes, characters, settings, or events, making the reading experience more vivid and engaging. By forming mental images, you can better understand and remember details, make connections, and engage more deeply with the material. This strategy helps clarify abstract concepts, enhances recall, and fosters a more immersive reading experience, ultimately improving overall comprehension.

Applying the Strategies for Comprehension and Understanding

The following section demonstrates an approach that a test taker may use when reading the required ESG Certification exam texts and applying the 12 strategies for comprehension and understanding to an example text.

The selected example text used in this chapter is the **AICPA ESG Reporting and** Attestation Roadmap 2021.[16] The test taker is encouraged to download this text (https://thecaqprod.wpenginepowered.com/wpcontent/uploads/2021/02/caq-esg-reporting-andattestation-roadmap-2021-Feb_v2.pdf) and follow along as the 12 strategies are applied to this example text.

In brief, AICPA's ESG Reporting and Attestation Roadmap 2021 provides audit practitioners with actionable guidance for supporting companies in ESG reporting and attestation. It underscores the growing importance of ESG disclosures driven by investor demand, regulatory trends, and stakeholder expectations. The document outlines risks, such as legal liabilities from inaccurate ESG data, and emphasizes third-party attestation's role in enhancing report reliability. Key sections detail how to integrate ESG information into SEC submissions, structure attestation reports, and navigate the lack of standardized metrics.

For auditors, the roadmap highlights the need to adapt traditional audit skills to ESG contexts, prioritize materiality, and collaborate with legal counsel to mitigate risks. It concludes that high-quality ESG reporting is essential for corporate accountability, requiring practitioners to balance innovation with rigor in this emerging field.

Example Analysis: Strategies for Comprehension and Understanding

The 12 strategies for reading comprehension and understanding, listed, have been applied to the example text.

1. **Ask Questions**
   Example: When reading about ESG reporting drivers, ask: "Why have companies turned to ESG reporting?" This prompts critical engagement with the link between ESG risks, meeting the information needs of their stakeholders, and providing transparency about commitments to identifying, managing, and reporting on ESG risks. Or when reading about ESG reporting frameworks (SASB, GRI, TCFD), ask: "How do these frameworks differ, and which aligns best with SEC requirements?"
2. **Connect with the Topic**
   Example: Relate the roadmap's discussion of ESG attestation to prior knowledge of financial audits. For instance, "ESG attestation mirrors financial audits but focuses on sustainability metrics – how might my audit skills transfer here?"
3. **Draw Conclusions**
   Example: The roadmap notes that "third-party assurance from an independent accounting firm can enhance the reliability of ESG information reported by companies." Infer: "Independent assessment and reporting likely reduces investor risk perception, similar to audits of financial statements and internal control over financial reporting."
4. **Engage in Discussions**
   Example: Debate with peers: "The roadmap emphasizes third-party assurance – should ESG attestation be mandatory like financial audits or voluntary?"

5. **Hone Accurate Reading Skills**
   Example: Practice fluency by rereading complex sections, such as "Practitioner Risk and Legal Considerations," to grasp the legal implications of ESG attestation errors.
6. **Identify Main Idea**
   Example: The main theme is clear: "Third-party assurance from an independent accounting firm can enhance the reliability of ESG information reported by companies, address *independent assurance to meet stakeholder demands, and is similar to* the process that occurs with audits of financial statements and internal control over financial reporting." The objective of the text, as stated within, is, "The purpose of this roadmap is to assist practitioners in supporting companies with their ESG reporting goals."
7. **Predict a Direction**
   Example: After reading about users of ESG information, including investors and other stakeholders, who are seeking transparency about ESG initiatives, and also demanding high-quality, accurate, reliable information. Predict: "the percentage of organizations seeing value in having some or all of the information in disclosure of ESG information in SEC submissions, being subject to external attestation per attestation standards (e.g., AICPA attestation standards) by an independent practitioner."
   After reviewing regulatory trends, infer: "Standardized ESG metrics will likely emerge, increasing attestation demands."
8. **Preview the Material**
   Example: Skim headings to note the roadmap's focus on "risk, legal considerations, and reporting frameworks" before deep reading. The roadmap begins with a clear purpose statement and headers (e.g., "Risk and Legal Considerations"), allowing you to predict content focus.
9. **Summarize/Paraphrase/Synthesize**
   Summarize the material you have read, "in your own words." Example: The chapter "Practitioner risk and legal considerations," focuses on the critical risk and legal considerations that practitioners must address when providing attestation services for ESG (Environmental, Social, and Governance) information, especially when this information is included in SEC filings. It emphasizes the importance of adhering to AICPA's Attestation Standards and Quality Control Standards throughout the engagement, from client acceptance to reporting.
   Key decision attributes include defining the scope of the attestation, determining suitable reporting criteria (e.g., GRI, SASB, company-developed metrics), selecting the level of attestation service (examination or review engagement), and understanding how the ESG information

and attestation report will be disclosed and used. The document highlights that the level of attestation service (examination or review) should align with the company's objectives, materiality considerations, intended messaging, and the needs and expectations of users.

10. **Think about Thinking (Metacognition)**
    Example: Reflect while reading, "How does my experience with financial audits apply to ESG attestation? What new skills do I need?" How can I use my current skill set to better understand ESG attestation requirements? What type of multiple-choice exam questions can come from this material?
11. **Utilize Background Knowledge**
    Example: Use familiarity with the audit process and GAAP to understand ESG frameworks like GRI/SASB and how they differ in measuring non-financial data. The document links ESG attestation to familiar audit processes. Third-party assurance from an independent accounting firm can enhance the reliability of ESG information reported by companies, like the process that occurs with audits of financial statements and internal control over financial reporting.
12. **Visualize**
    Example: Imagine a company's ESG report with sections for carbon emissions (E), diversity stats (S), and board structure (G) to contextualize the attestation information provided within the roadmap's guidance.

Reader Takeaways from Example Text

As you read any text, especially one in preparation for an exam, determine the main theme and the text's objective. For the example text used in this section.

1. Main Theme
   ESG reporting and third-party attestation are critical for transparent, reliable sustainability disclosures, requiring careful risk management by audit practitioners.
2. Objective
   To guide auditors in navigating ESG reporting complexities, including legal risks, attestation best practices, and stakeholder communication.

Identify keywords used throughout the text. Look up the definition of unfamiliar terms. In the example text, for example:

- ESG Reporting: Disclosures on Environmental, Social, and Governance metrics.
- Attestation: Independent verification of ESG data.

- Materiality Principle: Reporting ESG factors that significantly impact stakeholders.
- Examination Engagement: High-level assurance (similar to audits) vs. Review Engagement: Limited assurance.

Enumerate the most important takeaways you have gained from reading the text. From the example document:

- ESG reporting is evolving rapidly, driven by investor demand.
- Third-party assurance boosts credibility.
- Legal risks (e.g., liability for inaccurate ESG data) require careful mitigation.
- ESG reporting lacks standardization, increasing reliance on frameworks like SASB and GRI.

Summarized the new knowledge gained or existing knowledge reinforced upon completing the text. For example, from the AICPA text:

- ESG reporting frameworks (e.g., GRI, SASB) are not yet standardized, creating challenges for auditors.
- Attestation processes for ESG data differ from financial audits but share core principles of objectivity and rigor.
- ESG metrics can be company-specific or framework-aligned, affecting attestation complexity.
- Investors prioritize comparability, driving demand for third-party assurance.

SUMMARY

This chapter presents a systematic exploration of reading comprehension – the ability to grasp and integrate meaning from a text – and reading for understanding, which emphasizes decoding and identifying the literal meaning of content, emphasizing their importance as distinct yet interconnected processes.

Several theoretical models/frameworks that inform reading development are introduced which provide foundational insight into how individuals process text and the various components that contribute to successful reading and comprehension.

The chapter outlines 12 practical strategies for enhancing reading comprehension and understanding and applies these strategies through an example analysis of the AICPA ESG Reporting and Attestation Roadmap, demonstrating how each technique supports deeper engagement with technical content. This approach empowers test takers to actively process, retain, and apply information effectively.

Chapter 3 emphasizes that reading is not a passive activity but an active cognitive process that requires strategic engagement. By applying the principles and strategies presented, test takers can improve overall reading skills, which are particularly relevant in preparation for professional assessment exams and can increase a broader capacity for critical thinking and knowledge acquisition.

TEST YOURSELF

The following five multiple-choice questions are developed from the example text, "AICPA ESG Reporting and Attestation Roadmap 2021." Test your improved reading comprehension and understanding skills by downloading and reading the full document, then answer the sample questions. Turn to Appendix 1 to check your answer.

Question 1

According to the AICPA ESG Reporting and Attestation Roadmap 2021, what is the role of independent auditors in the flow of reliable information?

a. To create ESG reports for companies.
b. To make investment decisions for stakeholders.
c. To enhance the reliability of ESG information reported by companies.
d. To determine which ESG information should be reported.

Question 2

Which of the following is considered a key reason for the expansion of ESG reporting?

a. Requirements from the Internal Revenue Service.
b. Increasing interest from investors in ESG transparency and disclosure.
c. Changes in Generally Accepted Accounting Principles (GAAP).
d. Reduction in company tax obligations.

Question 3

What is the objective of a review-level attestation engagement according to AICPA standards?

a. To obtain limited assurance and express a conclusion about material modifications.
b. To provide absolute assurance about financial statements.
c. To determine tax liabilities of ESG investments.
d. To perform a full audit of all financial controls.

Question 4
Why might company-developed ESG metrics present challenges for attestation engagements?

a. They are always more accurate than standardized metrics
b. They may lack comparability and make criteria evaluation more difficult
c. They are legally required for all public companies
d. They are prohibited by AICPA standards

Question 5
What percentage of S&P 500 companies subjected some or all of their sustainability information to some sort of third-party assurance in 2019?

a. 10%
b. 29%
c. 50%
d. 75%

NOTES

1 Fielding, L. and Pearson, D., February 1994, Reading Comprehension: What Works, *Educational Leadership*, 51, 5, 62–68, www.researchgate.net/publication/254343574_Reading_comprehension_What_works.
2 NAAL, 2003, "The 2003 National Assessment of Adult Literacy," National Assessment of Adult Literacy, and National Center for Education Statistics (NCES), https://nces.ed.gov/naal/kf_demographics.asp#1.
3 PIAAC, 2023, "Highlights of the 2023 U.S. PIAAC Results," Program for the International Assessment of Adult Competencies, https://nces.ed.gov/surveys/piaac/2023/national_results.asp.
4 The National Assessment of Educational Progress (NAEP), 2025, "Reading Framework For The 2026 National Assessment of Educational Progress," U.S. Department of Education. Developed for the National Assessment Governing Board under contract number 91995918C0001 by WestEd, with a subcontract to the Council of Chief State School Officers, page 11. www.nagb.gov/content/dam/nagb/en/documents/publications/frameworks/reading/2026-reading-framework/naep-2026-reading-framework.pdf.
5 Marcella, A., 2025, Figure 3.1, A Heuristic For Thinking About Reading Comprehension, RAND Corporation, MR-1465-OERI, redrawn by authors from Snow, C.E., 2002, "Reading for Understanding: Toward an R&D Program in Reading Comprehension," RAND Corporation, MR-1465-OERI, www.rand.org/pubs/monograph_reports/MR1465.html.

6 Snow, C.E., 2002, "Reading for Understanding: Toward an R&D Program in Reading Comprehension," RAND Corporation, MR-1465-OERI, www.rand.org/pubs/monograph_reports/MR1465.html.
7 Marcella, A, 2025, Figure 3.2, The Components of Comprehension, created by authors.
8 Gough, P. and Tunmer, W., 1986, Decoding, Reading, and Reading Disability. *Remedial and Special Education*, 7, 6–10, https://citeseerx.ist.psu.edu/document?repid=rep1&type=pdf&doi=ee3c58058290e1ff077c96fa87f8fc8775d49b33#:~:text=Decoding%2C%20Reading%2C%20and-,Reading%20Disability,or%20garden%20variety%2C%20reading%20disability.
9 Marcella, A., 2025, Figure 3.3, Simple View of Reading, created by authors.
10 Zimmermann, S., July 15, 2003, *Seven Keys To Comprehension: How to Help Your Kids Read It and Get It!*, Clarkson Potter/Ten Speed.
11 Interactive English Language Centre, 2022, "Narrative Text – Definition, Characteristics, Functions, Types, and Example," https://bookishbay.com/narrative-text-definition.
12 (n.a.), 2025, *Comprehension: The Goal of Reading*, Read Naturally, Inc., www.readnaturally.com/research/5-components-of-reading/comprehension.
13 Nowak, P., August 16, 2022, "What are the Benefits of Reading Comprehension?" Iris Reading, LLC, https://irisreading.com/what-are-the-benefits-of-reading-comprehension.
14 Sedita, J., October 8, 2024, "Making Inferences to Support Comprehension," Keys to Literacy, https://keystoliteracy.com/blog/making-inferences-to-support-comprehension.
15 (n.a.), June 3, 2024, "Empowering Readers: Tools and Techniques for Improving Reading Comprehension," Neuhaus Education Center, https://neuhaus.org/empowering-readers-tools-and-techniques-for-improving-reading-comprehension.
16 AICPA, 2021, "AICPA ESG Reporting-and-attestation-roadmap-2021," The Association of International Certified Professional Accountants and The Chartered Institute of Management Accountants (CIMA), https://thecaqprod.wpenginepowered.com/wpcontent/uploads/2021/02/caq-esg-reporting-andattestation-roadmap-2021-Feb_v2.pdf.

REFERENCES

This chapter was written and organized, utilizing but not limited to the following additional sources. Custom modification to content adapted from these sources was done to present material more closely and accurately representative of the examination for which the test candidate is preparing to take.

Amplify Staff, June 27, 2024, "Reading Comprehension Strategies Grounded in Science," Amplify, https://amplify.com/blog/science-of-reading/reading-comprehension-strategies-grounded-in-science.

Andsager, G., November 26, 2024, "Reading Comprehension Basics: Types and Strategies," 95 Percent Group LLC, www.95percentgroup.com/insights/reading-comprehension-basics-types-and-strategies.

Cooley, M., October 23, 2024, "7 High Impact, Evidence-based Reading Comprehension Strategies (Processes) for Educators," 95 Percent Group LLC, www.95percentgroup.com/insights/7-high-impact-evidence-based-reading-comprehension-strategies-processes-for-educators.

Faulconbridge, R.I., January 1, 2008, "The Development of Learning and Teaching Strategies and Technical Texts for Diverse Groups of Adult Learners," Doctoral Thesis, University of Southern Queensland, https://research.usq.edu.au/download/6964e39025b8ecbe52dcd5356ad253a721f5b98138be547ddf90c444627 87b1b/17482964/Faulconbridge_2008_whole.pdf.

Hoover, W. A. and Gough, P.B., June 1990, The Simple View of Reading. *Reading and Writing: An Interdisciplinary Journal*, 2, 127–160. https://dx.doi.org/10.1007/BF00401799.

IMSE, September 1, 2023, "What Is Comprehension? Part 6 of The Essential Components of Literacy," Institute for Multi-Sensory Education, IMSE Journal, https://journal.imse.com/what-is-comprehension/#:~:text=Comprehension%20is%20the%20very%20impetus,text%20involving%20thinking%20and%20reasoning.

Lexia, November 18, 2024, "What is Scarborough's Reading Rope?," Lexia Blog, www.lexialearning.com/blog/what-is-scarboroughs-reading-rope

National Institute of Child Health and Human Development, 2000, "Report of the National Reading Panel: Teaching Children to Read," Washington, DC, www.nichd.nih.gov/sites/default/files/publications/pubs/nrp/Documents/report.pdf.

Oakhill, J., Cain K., and Elbro, C., August 20, 2014, *Understanding and Teaching Reading Comprehension: A Handbook*, Routledge.

Sadikovna, Z. and Sarvinoz Ahadjon qizi, E., n.d., "Developing Reading Skills: Strategies for Effective Reading," Global Trends in Education and Research Development, https://gisconf.com/index.php/GTERD/article/download/453/466.

Snow, C.E., 2002, "Reading for Understanding: Toward an R&D Program in Reading Comprehension," RAND Corporation, MR-1465-OERI, www.rand.org/pubs/monograph_reports/MR1465.html.

Talwar, A., Fall 2024, "The Science of Reading and Where It Stands in Adult Education, American Institutes for Research," American Institutes for Research, https://doi.org/10.35847/ATalwar.6.3.43.

Terry, B., January 21, 2025, "10 Best Practices to Improve Reading Comprehension," https://scholarwithin.com/10-best-practices-to-improve-reading-comprehension

Voyager Sopris Learning, September 15, 2023, "Comprehension: Defining the Key to Understanding," www.voyagersopris.com/vsl/blog/comprehension-defining-the-key-to-understanding.

Willingham, D., December 1, 2023, Beyond Comprehension, *ASCD*, 81(4), www.ascd.org/el/articles/beyond-comprehension.

Zimmerman, S. and Hutchins, C., 2003, *Seven Keys to Comprehension: How to Help Your Kids Read It and Get It!* New York: Three Rivers Press.

Chapter 4

Executive-Level Sample Questions

QUESTION 1

According to Global Reporting Initiative (GRI) Standard GRI 2, effective January 2023, Section 3 – Governance – covers disclosure that provides information about the organization's governance structure, composition, knowledge, roles, and remuneration. Requirements relevant to the highest governance body within the Governance structure and composition include which of the following three?

A. Describe its governance structure, describe the nomination and election process criteria, and describe the due diligence process.
B. Describe the highest governance bodies' roles and responsibilities, detail due diligence in the selection process and ensure free from conflicts of interest.
C. Describe its governance structure, list the committees of the highest governance body, and describe the composition of the highest governance body and its committees.
D. Describe which governing body committees oversee management, list committee members' competencies and state the impact the governing body has on the economy.

DOI: 10.1201/9781032672557-4

QUESTION 2

As identified in the Paris Agreement, what two specific areas address technology development and transfer?

A. To adapt to climate change impacts and foster climate resilience.
B. Develop nationally determined contributions and national adaptation plans.
C. Establish low-carbon solutions and open new markets.
D. Improve resilience to climate change and reduce greenhouse gas (GHG) emissions.

QUESTION 3

An ESG Executive is evaluating the adoption of ISSB standards within their multinational corporation. Considering the article "IFRS Sustainability-related risks and opportunities and the disclosure of material information," discussion on interoperability, which of the following statements best reflects a strategic approach to aligning with other sustainability reporting frameworks like ESRS and GRI?

A. Prioritize full compliance with ESRS and GRI standards first, as these frameworks offer broader stakeholder coverage than ISSB.
B. Focus solely on meeting the requirements of ISSB standards, with minimal consideration for alignment with other frameworks to avoid complexity.
C. Recognize that ESRS uses the term materiality to both identify the "material" matters and their related impacts, risks, and opportunities to report on and to assess the "materiality of information" disclosed on those matters.
D. Develop a dual reporting system that strictly separates ISSB disclosures for investors from ESRS/GRI disclosures for other stakeholders, ensuring no overlap.

QUESTION 4

According to SASB Guidance, if an entity omits one or more disclosure topics and/or accounting metrics, the entity should disclose the omission(s) along with what other information.

A. The rationale for the omission/s.
B. The lack of applicability.
C. The lack of associated metrics.
D. The standards are too new to enforce.

QUESTION 5

Principle 10 of the United Nations Global Compact (UNGC) states that businesses should work against corruption in all its forms to combat corruption and unethical behavior and specifically addresses which two actions?

A. Bribery and mismanagement.
B. Extortion and bribery.
C. Mismanagement and fraud.
D. Extortion or kickbacks.

QUESTION 6

Despite the green appetite of power consumers, including AI, according to the article "AI's Energy Footprint Warrants Markets, Not Panic," what is the primary reason that prevents suppliers from developing innovative products that best reduce the indirect emissions of power consumption?

A. The density of AI start-up ecosystems in major metropolitan areas.
B. Unclear and overreaching financial regulation.
C. A lack of government subsidies for AI-driven projects.
D. The overall size of the AI data center industry.

QUESTION 7

According to ISO 26000, compliance with the law is a fundamental duty of any organization and should be an essential part of what internal program within an organization?

A. Social responsibility.
B. Fair labor practices.
C. Right to work programs.
D. A safe work environment.

QUESTION 8

Considering the evolving ESG landscape, as highlighted in the "IFC ESG Guidebook," what strategic approach should an ESG Executive prioritize to ensure long-term value creation for their organization?

A. Linking environmental, social, and governance factors to the overall conduct and performance of the organization.
B. Focusing solely on short-term profitability while addressing ESG concerns reactively to minimize immediate costs.
C. Prioritizing environmental aspects of ESG while neglecting social and governance considerations to demonstrate a commitment to sustainability.
D. Limiting ESG integration to compliance-driven initiatives to avoid unnecessary investments and maintain operational efficiency.

QUESTION 9

According to the Sustainability Development Goals Report for 2023, this sector is a critical driver of productivity, employment, and growth, as business leaders are increasingly acknowledging the necessity and urgency of taking sustainability factors into account to achieve long-term success.

A. Government.
B. Technology.
C. Non-profits.
D. Private.

QUESTION 10

According to the SEC's final rules document, "The Enhancement and Standardization of Climate-Related Disclosures for Investors," which financial filings will now require climate-related information?

A. Form 10-Q.
B. Annual reports.
C. Schedule 14D-9.
D. Registration statements and annual reports.

QUESTION 11

According to the World Bank's Environmental and Social Framework (ESF), where remedy is not possible, the Borrower will, within a reasonable period, shift the project's primary suppliers to suppliers that can demonstrate that they are meeting the relevant requirements of the Environmental and Social Standards (ESS). The ability to address supplier risks is often dependent on what?

A. The level of control or influence primary suppliers hold over the Borrower.
B. The level of control or influence the Borrower holds over its primary suppliers.
C. The level of control or influence shareholders have over the Borrower.
D. The level of control through the World Bank over the Borrower and their suppliers.

QUESTION 12

Which of the following is a key responsibility of an organization's board of directors under IFRS S1 and S2 regarding sustainability disclosures?

A. Delegating sustainability tasks to specialized departments and monitoring their progress.
B. Assessing the climate, environmental, and social risks and opportunities identified as material to the business.
C. Approving the annual sustainability report prepared by the sustainability department.
D. Ensuring sustainability-related financial information is disclosed separately from financial statements.

QUESTION 13

According to ISO 14001, what two activities, when implemented, will assist the organization in improving environmental performance, gaining both a competitive advantage and the trust of stakeholders?

A. Clearly documented policies and procedures.
B. Staff engagement programs and community activities.
C. More efficient use of resources and waste reduction.
D. A corporate-wide recycle and reuse program to eliminate waste.

QUESTION 14

According to the OECD Guidelines for Multinational Enterprises on Responsible Business Conduct, how should organizations address adverse impacts that are directly linked to their operations, products, or services through a business relationship?

Organizations should:

A. Terminate the relationship immediately to avoid liability.
B. Document and report impacts caused by third parties in their supply chain.
C. Publicly disclose all business relationships to ensure transparency, regardless of risk.
D. Seek to prevent or mitigate the impact through leverage, collaboration, or other influence strategies.

QUESTION 15

According to the United Nations Guiding Principles on Business and Human Rights, to gauge human rights risks, business enterprises should identify and assess any actual or potential adverse human rights impacts with which they may be involved either through their own activities or as a result of their business relationships. Which two key points should this process include?

A. Internal and/or independent external human rights expertise and enterprise management and its highest governing body.
B. Enterprise management only and the potentially affected groups and other relevant stakeholders.
C. Internal and/or independent external human rights expertise and the potentially affected groups and other relevant stakeholders.
D. External third-party consultants with relevant human rights experience and potentially affected groups and other relevant stakeholders.

Executive-level Sample Answers are found in Chapter 6.

Chapter 5

Generalist-Level Sample Questions

QUESTION 1

While recognizing and following international norms of behavior, applying ISO 26000 requires an organization to also consider:

A. Labor contracts.
B. Management targets.
C. Due diligence processes.
D. Economic conditions.

DOI: 10.1201/9781032672557-5

QUESTION 2

The term __________ is used throughout the SASB guidance standards to reflect requirements of the standards, while the terms __________ and __________ are used to indicate guidance, which, although not required and optional, provides a recommended approach for the implementation of the standard.

A. Will, could, possibly.
B. Must, should, could.
C. Shall, should, and may.
D. Required, could, perhaps.

QUESTION 3

The main purpose of IFRS S1 and S2 standards is to:

A. Identify possible environmental and social risks, regardless of their impact on the company.
B. Meet investors' desire for information that is reliable, comparable, and connected to the financial statements.
C. Prioritize the organization's philanthropic activities to improve the organization's public image.
D. Identify and address biodiversity loss and ecosystem degradation.

QUESTION 4

According to ISO 14001, what single benefit provides a strategic approach to improving an organization's environmental performance?

A. Illustrating an increase in market share.
B. Meeting regulatory or contractual requirements.
C. Establishing a strategic planning process.
D. Demonstrating strong leadership and board.

QUESTION 5

According to the document, "IFRS Sustainability-related risks and opportunities and the disclosure of material information," when identifying sustainability-related risks and opportunities, which of the following approaches aligns with the ISSB Standards?

A. Focus exclusively on risks and opportunities that have immediate financial impacts.
B. Consider only the sustainability-related risks and opportunities identified by industry peers.
C. Identify sustainability-related risks and opportunities across the entity's value chain.
D. Limit the assessment to environmental risks, as they are the most quantifiable.

QUESTION 6

What type of information regarding greenhouse gas (GHG) emissions will certain larger companies be required to disclose under the SEC's final rules, outlined in the document "The Enhancement and Standardization of Climate-Related Disclosures for Investors"?

A. Scope 3 emissions, as they are significant to maintaining a zero carbon footprint.
B. Scope 1 emissions, as they are directly controlled by the company.
C. Scope 1 and/or Scope 2 emissions, regardless of materiality.
D. Scope 1 and/or Scope 2 emissions, when those emissions are material.

QUESTION 7

According to Global Reporting Initiative (GRI) Standard GRI 2, effective January 2023, Section 1 – Key Concepts – the organization prioritizes reporting on those topics that represent its most significant impacts on the economy, environment, and people, including impacts on their human rights. The topics in the report are as:

A. Impact statement topics.
B. Sustainability topics.
C. Material topics.
D. Value creation topics.

QUESTION 8

According to the Sustainability Development Goals Report for 2023, globally, women accounted for nearly 40% of total employment but held only 28.2% of management positions in 2021. Progress on __________ and increasing women's representation in management has been slow, with representation in management below their share in total employment across all regions.

A. Gender parity.
B. Managerial positions.
C. Gender quotas.
D. Legislated rights.

QUESTION 9

The World Bank's Environmental and Social Framework (ESF) states that all parties who employ or engage project workers will develop and implement:

A. GHG levels to meet UN 2025 goals.
B. A safe working environment.
C. A safety training program.
D. Universal collective bargaining policies.

QUESTION 10

What is the primary role of National Contact Points (NCPs) under the OECD Guidelines for Multinational Enterprises on Responsible Business Conduct? The primary role of the Guidelines is to:

A. Handle complaints, facilitate dialogue, and offer guidance on responsible business conduct.
B. Act as an enforcement body that penalizes companies for non-compliance with the Guidelines.
C. Provide financial assistance to multinational enterprises implementing responsible business practices.
D. Develop country-specific laws that align with the OECD Guidelines.

QUESTION 11

According to Principle 7 of the United Nations Global Compact (UNGC), businesses should support what type of approach to environmental challenges and apply this when there is reasonable suspicion of harm.

A. Precautionary.
B. Visionary.
C. Responsible.
D. Methodology.

QUESTION 12

The Paris Agreement established an enhanced transparency framework (ETF), for subscribing countries to report actions taken and progress in climate change mitigation, adaptation measures and support provided or received. The information gathered through the ETF generates which specific document?

A. Transparency report.
B. Progress report.
C. Global stocktake report.
D. Climate change report.

QUESTION 13

The document, "AI's Energy Footprint Warrants Markets, Not Panic," mentions "co-location" as a practice reducing infrastructure burdens and emissions from data center expansion. What does "co-location" refer to in this context?

A. Data centers built out in geographic clusters, prioritizing network connectivity.
B. Data centers are requesting capacity in multiple utility footprints to expand a single data center.
C. Co-location reduces flexibility to locate demand where it is least burdensome to infrastructure.
D. Tech firms collaborate with competitive power plant owners to build data centers at the site of nuclear plants.

QUESTION 14

According to the "IFC ESG Guidebook," what are the three core areas of integration within the ESG Integration Framework?

A. Governance, Management, and Disclosure
B. Philanthropy, Volunteering, and Community Engagement
C. Environment, Social, and Technology
D. Compliance, Reporting, and Auditing

QUESTION 15

According to the United Nations Guiding Principles on Business and Human Rights, to meet their responsibility to respect human rights, business enterprises should have in place policies and processes appropriate to their size and circumstances, including which of the following?

A. A due diligence process to identify, prevent, mitigate, and account for their impact on human rights.
B. A human rights policy, a due diligence process to address their impact on human rights, and a remediation process.
C. A commitment to a human rights policy and a remediation process for potentially affected groups.
D. A committee including the human rights chief officer and Board committee member, and a representative of the affected group.

Generalist-level Sample Answers are found in Chapter 6.

Chapter 6

Executive-Level and Generalist-Level Sample Questions

Correct Answers and Substantiation

ANSWERS TO EXECUTIVE-LEVEL SAMPLE QUESTIONS

Question 1

According to the Global Reporting Initiative (GRI) Standard GRI 2, effective January 2023, Section 3 – Governance – covers disclosure that provides information about the organization's governance structure, composition, knowledge, roles, and remuneration. Requirements relevant to the highest governance body within the Governance structure and composition include which of the following three?

C. Describe its governance structure, list the committees of the highest governance body, and describe the composition of the highest governance body and its committees.

Source: (n.a.), October 5, 2021, "General Disclosures 2021," GRI2_General Disclosures2021.pdf, page 20.

❋

Question 2

As identified in the Paris Agreement, what two specific areas address technology development and transfer?

D. Improve resilience to climate change and reduce greenhouse gas (GHG) emissions.

Source: (n.a.), (n.d.), "What is the Paris Agreement?," United Nations, https://unfccc.int/process-and-meetings/the-paris-agreement/the-paris-agreement.

❋

DOI: 10.1201/9781032672557-6

Question 3

An ESG Executive is evaluating the adoption of ISSB standards within their multinational corporation. Considering the article "IFRS Sustainability-related risks and opportunities and the disclosure of material information," discussion on interoperability, which of the following statements best reflects a strategic approach to aligning with other sustainability reporting frameworks like ESRS and GRI?

C. Recognize that ESRS uses the term materiality to both identify the "material" matters and their related impacts, risks, and opportunities to report on and to assess the "materiality of information" disclosed on those matters.

Source: ISSB, November 2024, "IFRS Sustainability-related Risks and Opportunities and the Disclosure of Material Information," International Sustainability Standards Board, www.ifrs.org/content/dam/ifrs/supporting-implementation/issb-standards/issb-materiality-education-material.pdf, page 6.

❋

Question 4

According to SASB Guidance, if an entity omits one or more disclosure topics and/or accounting metric, the entity should disclose the omission(s) along with what other information.

A. The rationale for the omission/s.

Source: (n.a.), 2022, "SASB Standards Application Guidance," The IFRS Foundation, SASB-Standards-Application-Guidance-2018-10.pdf, page 2.

❋

Question 5

Principle 10 of the United Nations Global Compact (UNGC) states that businesses should work against corruption in all its forms to combat corruption and unethical behavior and specifically addresses which two actions?

B. Extortion and bribery.

Source: (n.a.), (n.d.), "The Ten Principles of the UN Global Compact," https://www.unglobalcompact.org/what-is-gc/mission/principles.

❋

Question 6

Despite the green appetite of power consumers, including AI, according to the article "AI's Energy Footprint Warrants Markets, Not Panic," what is the primary reason that prevents suppliers from developing innovative products that best reduce the indirect emissions of power consumption?

B. Unclear and overreaching financial regulation.

Source: Hartman, D. and Manzagol, O., September 26, 2024, "AI's Energy Footprint Warrants Markets, Not Panic," R Street Institute, www.rstreet.org/commentary/ais-energy-footprint-warrants-markets-not-panic, page 10.

❋

Question 7

According to ISO 26000, compliance with the law is a fundamental duty of any organization and should be an essential part of what internal program within an organization?

A. Social responsibility.

Source: (n.a.), 2018, "Discovering ISO 26000," Discovering ISO 26000 – Guidance on social responsibility, page 6.

❋

Question 8

Considering the evolving ESG landscape, as highlighted in the "IFC ESG Guidebook," what strategic approach should an ESG Executive prioritize to ensure long-term value creation for their organization?

A. Linking Environmental, Social, and Governance factors to the overall conduct and performance of the organization.

Source: Delange, N., et al., 2021, "IFC ESG Guidebook," International Finance Corporation, www.ifc.org/content/dam/ifc/doc/mgrt/ifc-esg-guidebook.pdf, page 50

❋

Question 9

According to the Sustainability Development Goals Report for 2023, this sector is a critical driver of productivity, employment and growth, as business leaders are increasingly acknowledging the necessity and urgency of taking sustainability factors into account to achieve long-term success.

D. Private.

Source: (n.a.), July 2023, "The Sustainable Development Goals Report 2023, Special Edition," United Nations System, https://unstats.un.org/sdgs/report/2023/The-Sustainable-Development-Goals-Report-2023.pdf, page 49.

❋

Question 10

According to the SEC's final rules document, "The Enhancement and Standardization of Climate-Related Disclosures for Investors," which financial filings will now require climate-related information?

D. Registration statements and annual reports.

Source: SEC, May 28, 2024, "The Enhancement and Standardization of Climate-Related Disclosures for Investors," Securities and Exchange Commission, www.sec.gov/files/rules/final/2024/33-11275.pdf, page 1.

❋

Question 11

According to the World Bank's Environmental and Social Framework (ESF), where remedy is not possible, the Borrower will, within a reasonable period, shift the project's primary suppliers to suppliers that can demonstrate that they are meeting the relevant requirements of the Environmental and Social Standards (ESS). The ability to address supplier risks is often dependent on what?

B. The level of control or influence the Borrower holds over its primary suppliers.

Source: (n.a.), 2017, "The World Bank Environmental and Social Framework," ESFFramework.pdf, page 36.

❋

Question 12

Which of the following is a key responsibility of an organization's board of directors under IFRS S1 and S2 regarding sustainability disclosures?

B. Assessing the climate, environmental, and social risks and opportunities identified as material to the business.

Source: Instituto Brasileiro de Governança Corporativa, "Guide for Board Members: IFRS S1 and S2 Sustainability Standards," Instituto Brasileiro de Governança Corporativa, Chapter Zero Brazil, 55p.; PDF; 2.70 MB, guia_conselheiros_ING_P3.pdf, page 8.

❋

Question 13

According to ISO 14001, what two activities, when implemented, will assist the organization in improving environmental performance, gaining both a competitive advantage and the trust of stakeholders?

C. More efficient use of resources and waste reduction.

Source: (n.a.), 2015, "ISO 14001: Key Benefits," International Organization for Standardization (ISO). ISO – ISO 14001 – Key benefits, page 2.

❋

Question 14

According to the OECD Guidelines for Multinational Enterprises on Responsible Business Conduct, how should organizations address adverse impacts that are directly linked to their operations, products, or services through a business relationship?

Organizations should:

D. Seek to prevent or mitigate the impact through leverage, collaboration, or other influence strategies.

Source: OECD, June 8, 2023, "OECD Guidelines for Multinational Enterprises on Responsible Business Conduct," Organisation for Economic Co-operation and Development, OECD Publishing, Paris, www.oecd.org/content/dam/oecd/en/publications/reports/2023/06/oecd-guidelines-for-multinational-enterprises-on-responsible-business-conduct_a0b49990/81f92357-en.pdf, Section IV Human Rights, paragraph 47, page 26.

❋

Question 15

According to the United Nations Guiding Principles on Business and Human Rights, to gauge human rights risks, business enterprises should identify and assess any actual or potential adverse human rights impacts with which they may be involved either through their own activities or as a result of their business relationships. Which two key points should this process include?

C. Internal and/or independent external human rights expertise and the potentially affected groups and other relevant stakeholders.

Source: (n.a.), 2011, "Guiding Principles on Business and Human Rights: Implementing the United Nations 'Protect, Respect and Remedy' Framework," https://www.ohchr.org/documents/publications/guidingprinciplesbusinesshr_en.pdf, page 20.

****

ANSWERS TO GENERALIST-LEVEL QUESTIONS

Question 1

While recognizing and following international norms of behavior, applying ISO 26000 requires an organization to also consider:

D. Economic conditions.

Source: (n.a.), 2018, "Discovering ISO 26000," Discovering ISO 26000 – Guidance on social responsibility, page 7.

*

Question 2

The term __________ is used throughout the SASB guidance standards to reflect requirements of the standards, while the terms __________ and __________ are used to indicate guidance, which although not required and optional, provides a recommended approach for the implementation of the standard.

C. Shall, should, and may.

Source: (n.a.), 2022, "SASB Standards Application Guidance," The IFRS Foundation, SASB-Standards-Application-Guidance-2018-10.pdf, page 1.

*

Question 3

The main purpose of IFRS S1 and S2 standards is to:

B. Meet investors' desire for information that is reliable, comparable, and connected to the financial statements.

Source: Instituto Brasileiro de Governança Corporativa, "Guide for Board Members: IFRS S1 and S2 Sustainability Standards," Instituto Brasileiro de Governança Corporativa, Chapter Zero Brazil, 55p.; PDF; 2.70 MB, guia_conselheiros_ING_P3.pdf, page 16.

*

Question 4

According to ISO 14001, what single benefit provides a strategic approach to improving an organization's environmental performance?

B. Meeting regulatory or contractual requirements.

Source: (n.a.), 2015, "ISO 14001: Key Benefits," International Organization for Standardization (ISO), page 4.

*

Question 5

According to the document, "IFRS Sustainability-related risks and opportunities and the disclosure of material information," when identifying sustainability-related risks and opportunities, which of the following approaches aligns with the ISSB Standards?

C. Identify sustainability-related risks and opportunities across the entity's value chain.

Source: ISSB, November 2024, "IFRS Sustainability-related Risks and Opportunities and the Disclosure of Material Information," International Sustainability Standards Board, www.ifrs.org/content/dam/ifrs/supporting-implementation/issb-standards/issb-materiality-education-material.pdf, page 23.

*

Question 6

What type of information regarding greenhouse gas (GHG) emissions will certain larger companies be required to disclose under the SEC's final rules, outlined in the document "The Enhancement and Standardization of Climate-Related Disclosures for Investors?"

C. Scope 1 and/or Scope 2 emissions, when those emissions are material.

Source: SEC, May 28, 2024, "The Enhancement and Standardization of Climate-Related Disclosures for Investors," Securities and Exchange Commission, www.sec.gov/files/rules/final/2024/33-11275.pdf, page 15.

*

Question 7

According to Global Reporting Initiative (GRI) Standard GRI 2, effective January 2023, Section 1 – Key Concepts – the organization prioritizes reporting on those topics that represent its most significant impacts on the economy, environment, and people, including impacts on their human rights. The topics in the report are as:

D. Material topics.

Source: (n.a.), October 5, 2021, "General Disclosures 2021," GRI1_Foundation 2021.pdf, page 8.

❋

Question 8

According to the Sustainability Development Goals Report for 2023, globally, women accounted for nearly 40% of total employment but held only 28.2% of management positions in 2021. Progress on __________ and increasing women's representation in management has been slow, with representation in management below their share in total employment across all regions.

A. Gender parity.

Source: (n.a.), July 2023, "The Sustainable Development Goals Report 2023, Special Edition," United Nations System, https://unstats.un.org/sdgs/report/2023/The-Sustainable-Development-Goals-Report-2023.pdf, page 22.

❋

Question 9

The World Bank's Environmental and Social Framework (ESF) states that all parties who employ or engage project workers will develop and implement:

B. A safe working environment.

Source: (n.a.), 2017, "The World Bank Environmental and Social Framework," ESFFramework.pdf, page 34.

❋

Question 10

What is the primary role of National Contact Points (NCPs) under the OECD Guidelines for Multinational Enterprises on Responsible Business Conduct? The primary role of the Guidelines is to:

A. Handle complaints, facilitate dialogue, and offer guidance on responsible business conduct.

Source: OECD, June 8, 2023, "OECD Guidelines for Multinational Enterprises on Responsible Business Conduct," Organisation for Economic Co-operation and Development, OECD Publishing, Paris, www.oecd.org/content/dam/oecd/en/publications/reports/2023/06/oecd-guidelines-for-multinational-enterprises-on-responsible-business-conduct_a0b49990/81f92357-en.pdf, page 56.

❋

Question 11

According to Principle 7 of the United Nations Global Compact (UNGC), businesses should support what type of approach to environmental challenges and apply this when there is reasonable suspicion of harm.

A. Precautionary.

Source: (n.a.), (n.d.), "The Ten Principles of the UN Global Compact," https://www.unglobalcompact.org/what-is-gc/mission/principles.

❋

Question 12

The Paris Agreement established an enhanced transparency framework (ETF), for subscribing countries to report actions taken and progress in climate change mitigation, adaptation measures and support provided or received. The information gathered through the ETF generates which specific document?

C. Global stocktake report.

Source: (n.a.), (n.d.), "What is the Paris Agreement?," United Nations, https://unfccc.int/process-and-meetings/the-paris-agreement/the-paris-agreement.

❋

Question 13

The document, "AI's Energy Footprint Warrants Markets, Not Panic," mentions "co-location" as a practice reducing infrastructure burdens and emissions from data center expansion. What does "co-location" refer to in this context?

D. Tech firms collaborate with competitive power plant owners to build data centers at the site of nuclear plants.

Source: Hartman, D. and Manzagol, O., September 26, 2024, "AI's Energy Footprint Warrants Markets, Not Panic," R Street Institute, www.rstreet.org/commentary/ais-energy-footprint-warrants-markets-not-panic, page 9.

❋

Question 14

According to the "IFC ESG Guidebook," what are the three core areas of integration within the ESG Integration Framework?

A. Governance, Management, and Disclosure.

Source: Delange, N., et al., 2021, "IFC ESG Guidebook," International Finance Corporation, www.ifc.org/content/dam/ifc/doc/mgrt/ifc-esg-guidebook.pdf, page 11.

❋

Question 15

According to the United Nations Guiding Principles on Business and Human Rights, to meet their responsibility to respect human rights, business enterprises should have in place policies and processes appropriate to their size and circumstances, including which of the following?

B. A human rights policy, a due diligence process to address their impact on human rights and a remediation process.

Source: (n.a.), 2011, "Guiding Principles on Business and Human Rights: Implementing the United Nations 'Protect, Respect and Remedy' Framework," https://www.ohchr.org/documents/publications/guidingprinciplesbusinesshr_en.pdf, page 15.

❋❋❋❋

Appendix 1

Sample Practice Question Answers for Chapter 3: Reading for Comprehension and Understanding: Principles and Strategies

Sample text used to create example multiple-choice questions:

AICPA, 2021, "AICPA ESG Reporting-and-attestation-roadmap-2021," The Association of International Certified Professional Accountants and The Chartered Institute of Management Accountants (CIMA), https://thecaqprod.wpenginepowered.com/wpcontent/uploads/2021/02/caq-esg-reporting-andattestation-roadmap-2021-Feb_v2.pdf.

EXAMPLE MULTIPLE-CHOICE QUESTIONS

Question 1

According to the document, what is the role of independent auditors in the flow of reliable information?

a. To create ESG reports for companies.
b. To make investment decisions for stakeholders.
c. To enhance the reliability of ESG information reported by companies.
d. To determine which ESG information should be reported.

Correct Answer:

c. To enhance the reliability of ESG information reported by companies (Page 2).

Question 2

Which of the following is considered a key reason for the expansion of ESG reporting?

a. Requirements from the Internal Revenue Service.
b. Increasing interest from investors in ESG transparency and disclosure.
c. Changes in Generally Accepted Accounting Principles (GAAP).
d. Reduction in company tax obligations.

Correct Answer:

b. Increasing interest from investors in ESG transparency and disclosure (Page 4).

Question 3

What is the objective of a review-level attestation engagement according to AICPA standards?

a. To obtain limited assurance and express a conclusion about material modifications.
b. To provide absolute assurance about financial statements.
c. To determine tax liabilities of ESG investments.
d. To perform a full audit of all financial controls.

Correct Answer:

a. To obtain limited assurance and express a conclusion about material modifications (Page 10).

Question 4

Why might company-developed ESG metrics present challenges for attestation engagements?

a. They are always more accurate than standardized metrics.
b. They may lack comparability and make the criteria evaluation more difficult.
c. They are legally required for all public companies.
d. They are prohibited by AICPA standards.

Correct Answer:

b. They may lack comparability and make criteria evaluation more difficult (Page 10).

Question 5

What percentage of S&P 500 companies subjected some or all of their sustainability information to some sort of third-party assurance in 2019?

a. 10%
b. 29%
c. 50%
d. 75%

Correct Answer:

b. 29% (Page 5).

Appendix 2
ESG Certification Study Guide Resources

ENVIRONMENT

Sustainability Accounting Standards Board (SASB) Standards

Citation: (n.a.), 2022, "SASB Standards Application Guidance," The IFRS Foundation, SASB-Standards-Application-Guidance-2018-10.pdf.

Document Selection Substantiation

- **Global Influence:** SASB Standards are recognized internationally, with adoption by companies and investors seeking industry-specific ESG disclosures. SASB standards refer to, or incorporate by reference, laws, statutes, and third-party guidelines, protocols, standards, publications, and other references.
- **Regulatory Relevance:** SASB industry standards contain disclosure topics, associated accounting metrics and technical protocols, and activity metrics for each industry. They complement financial reporting and are increasingly integrated into regulatory frameworks.
- **Practical Applicability:** Offering industry-specific guidelines, they assist organizations in identifying material environmental issues pertinent to their sector. The SASB industry standards and the SASB Standards Application Guidance are subject to due process and governed by foundational documents: the SASB Conceptual Framework and the SASB Rules of Procedure.
- **Current Relevance:** SASB Standards are maintained under the IFRS Foundation, ensuring their alignment with current financial and sustainability reporting trends.

LEARNING OBJECTIVE

The reader should understand that SASB Standards Application Guidance (Version 2018-10) provides a comprehensive framework for entities to follow when applying SASB Sustainability Accounting Standards. The guidance emphasizes the voluntary nature of the standards and outlines the importance of reporting boundaries, data usage, transparency regarding omissions, modifications, and estimates. Additionally, it emphasizes governance, internal control, and potential third-party assurance for sustainability disclosures. This document is intended to ensure the accuracy and comparison of sustainability information reported by organizations.

AI's Energy Footprint Warrants Markets, Not Panic

Hartman, D., and Manzagol, O., September 26, 2024, "AI's Energy Footprint Warrants Markets, Not Panic," R Street Institute, www.rstreet.org/commentary/ais-energy-footprint-warrants-markets-not-panic.

Document Selection Substantiation

- **Global Influence:** The article suggests that AI's energy consumption is a global concern, advocating for market-driven solutions rather than panic-driven policies.
- **Regulatory Relevance:** It emphasizes the need for regulatory frameworks that balance AI's energy demands with grid reliability, promoting efficient energy use without stifling innovation.
- **Practical Applicability:** The piece proposes leveraging market mechanisms to address AI's energy footprint, encouraging industries to adopt sustainable practices through economic incentives.
- **Current Relevance:** Given the rapid advancement of AI technologies, the article underscores the urgency of implementing thoughtful policies to manage their energy impact effectively.

LEARNING OBJECTIVE

The reader should be able to analyze the impact of artificial intelligence (AI) on energy consumption, evaluate the implications of AI-driven electricity demand on market structures, and assess policy approaches to managing AI's energy footprint. Additionally, readers should understand the role of competitive electricity markets, regulatory frameworks, and infrastructure planning in addressing AI-related energy challenges while mitigating environmental and economic concerns.

United Nations Sustainable Development Goals (SDGs)

Citation: (n.a.), July 2023, "The Sustainable Development Goals Report 2023, Special Edition," United Nations System, https://unstats.un.org/sdgs/report/2023/The-Sustainable-Development-Goals-Report-2023.pdf.

Document Selection Substantiation

- **Global Influence:** The SDGs are a universal call to action to end poverty, protect the planet, and ensure prosperity for all, adopted by all United Nations Member States. It provides a comprehensive midpoint assessment of the 2030 Agenda, highlighting both areas of progress and those requiring acceleration.
- **Regulatory Relevance:** While not legally binding, the SDGs influence national policies and international agreements. The SDGs underscore that recent crises have disproportionately affected the world's poorest and most vulnerable populations.
- **Practical Applicability:** Organizations align their strategies and operations with the SDGs to contribute to global sustainability efforts. The report emphasizes the urgent need for strong political commitment and the effective use of available technologies, resources, and knowledge to accelerate progress toward the SDGs.
- **Current Relevance:** The SDGs remain a central framework for addressing environmental and social challenges globally. The report is prepared by the United Nations Department of Economic and Social Affairs (UN DESA) in collaboration with over 50 international and regional organizations, based on data from more than 200 countries and territories.

LEARNING OBJECTIVE

The Sustainable Development Goals (SDGs) Report 2023: Special Edition offers a comprehensive assessment of the 2030 Agenda for Sustainable Development. For the reader, it highlights challenges impeding progress, including the climate crisis, geopolitical conflicts, economic downturns, and the enduring effects of the COVID-19 pandemic. These emergencies have disproportionately affected vulnerable populations throughout the world. This report emphasizes the urgent need for robust commitment and the effective utilization of available technologies, resources, and knowledge to accelerate progress toward the Sustainable Development Goals.

The ESG Certification Exam will focus only on Goal 5 – General Equality and on Goal 7 – Affordable and Clean Energy.

ISO 14001: Key Benefits

Citation: (n.a.) 2015, "ISO 14001: Key Benefits," International Organization for Standardization (ISO). ISO – ISO 14001 – Key benefits.

Document Selection Substantiation

- **Global Influence:** ISO 14001 is widely adopted worldwide, for environmental management systems (EMS) with thousands of organizations using it to manage environmental impacts effectively. Top management must show leadership and commitment to the EMS by integrating it into the overall business strategy and decision-making processes.
- **Regulatory Relevance:** While voluntary, ISO 14001 helps organizations comply with environmental laws and regulations in various jurisdictions, to avoid violations and improve operational sustainability. It provides a path to ensure that proper resources, roles, and responsibilities are in place to carry out the EMS effectively, including training and awareness programs.
- **Practical Applicability:** It provides a structured framework for organizations to improve environmental performance through resource efficiency and waste reduction. Establishes processes to identify environmental impacts, assess risks, and create objectives, targets, and programs for minimizing environmental effects.
- **Current Relevance:** With rising regulatory and consumer focus on environmental responsibility, ISO 14001 remains highly relevant. ISO 14001 provides a path for maintaining clear records and documentation to support EMS processes, compliance, and continual improvement efforts.

LEARNING OBJECTIVE

The reader should understand ISO 14001 an international standard for environmental management systems (EMS). The standard provides a framework for organizations to follow to enhance environmental stewardship, comply with regulations, and continually improve their environmental impact. ISO 14001 applies to all types of organizations, regardless of size or industry, and emphasizes the importance of leadership commitment, systematic approaches to environmental management, and consistent improvement. It is designed to help organizations improve their environmental performance through more efficient use of resources and reduction of waste.

The Paris Agreement (United Nations Framework Convention on Climate Change) – "What is the Paris Agreement?"

Citation: (n.a.), (n.d.), "What is the Paris Agreement?," United Nations, https://unfccc.int/process-and-meetings/the-paris-agreement/the-paris-agreement.

Document Selection Substantiation

- **Global Influence:** The Paris Agreement is a legally binding international treaty on climate change, and landmark global accord aimed at limiting global warming to well below 2°C, with 194 signatories.
- **Regulatory Relevance:** It directly influences national environmental policies and commitments under Nationally Determined Contributions (NDCs), which encourages all Parties to put forward their best efforts and to strengthen the efforts in the years ahead.
- **Practical Applicability:** Guides organizations in aligning with climate goals and addressing carbon footprints. To make the Paris Agreement fully operational, a work program was launched to develop modalities, procedures, and guidelines on a broad array of issues.
- **Current Relevance:** As the primary framework for addressing climate change, the agreement remains at the center of global environmental governance. Since its introduction, low-carbon solutions and new markets have been developed, and more countries, regions, cities, and companies are establishing carbon neutrality targets.

LEARNING OBJECTIVE

The readers need to know that the Paris Agreement is an international treaty aimed at mitigating climate change by limiting the rise in global temperatures. Adopted in 2015, the agreement emphasizes national contributions to reduce greenhouse gas emissions, as well as global efforts to adapt to the impacts of climate change. It involves all countries, with each committing to nationally determined contributions (NDCs) that are periodically updated. The agreement also outlines mechanisms for financial support, transparency, and accountability to ensure global cooperation in tackling climate change.

SOCIAL

Global Reporting Initiative (GRI) Standards

Citation: (n.a.), October 5, 2021, "General Disclosures 2021," GRI2_GeneralDisclosures2021.pdf.

Document Selection Substantiation

- **Global Influence:** The GRI Standards are the world's most widely adopted sustainability reporting framework, utilized by organizations in over 100 countries.
- **Regulatory Relevance:** Many governments and regulatory bodies reference or mandate GRI Standards for corporate sustainability disclosures. The GRI Standards ensure that organizations can report on their sustainability journey in a structured, transparent, and standardized manner.
- **Practical Applicability:** They provide comprehensive guidelines for reporting on social topics such as labor practices, human rights, and community impacts.
- **Current Relevance:** Regular updates ensure alignment with evolving global sustainability challenges and stakeholder expectations.

LEARNING OBJECTIVE

The reader should understand that the **Global Reporting Initiative** (**GRI**) Standards are widely recognized guidelines for sustainability reporting. They aim to help organizations understand and communicate their economic, environmental, and social impacts, enabling better decision-making and transparency. The GRI Standards cover a broad range of topics, from climate change and human rights to governance and community well-being. They are structured to help organizations create reports that are relevant, consistent, and comparable, allowing stakeholders to assess the company's performance in a transparent and accountable way. The GRI Standards are designed to be adaptable to various types and sizes of organizations, ensuring that sustainability reporting is accessible and meaningful.

United Nations Global Compact (UNGC) Principles

Citation: (n.a.), (n.d.), "The Ten Principles of the UN Global Compact," https://www.unglobalcompact.org/what-is-gc/mission/principles.

Document Selection Substantiation

- **Global Influence:** The UNGC is the world's largest corporate sustainability initiative, helping businesses integrate responsible sustainability practices. Currently, there are over 15,000 participants across 160 countries.
- **Regulatory Relevance:** While voluntary, its principles align with international norms and influence national regulations and integrate responsible practices. By supporting the UN Global Compact

Principles, companies play a pivotal role in promoting ethical behavior and positive change within society and the global economy.

- **Practical Applicability:** The UNGC offers a framework for businesses to implement sustainable and socially responsible policies that foster trust, reduce risks, and contribute to broader sustainable development goals.
- **Current Relevance:** Continues to guide companies in aligning strategies with universal principles on human rights, labor, environment, and anti-corruption. The UNGC promotes ethical behavior and positive change within society and the global economy.

LEARNING OBJECTIVE

The reader should recognize that the United Nations Global Compact (UNGC) is a voluntary initiative for businesses and organizations to align their operations and strategies with universally accepted principles in the areas of human rights, labor, environment, and anti-corruption. It encourages companies to take action to meet sustainability goals and respect fundamental values, contributing to a more sustainable and inclusive global economy. By adhering to the ten UNGC principles, businesses commit to responsible corporate behavior, focusing on ethical standards that benefit both the business and society at large. The UNGC promotes transparency and accountability through regular reporting on progress made towards achieving these principles.

ISO 26000 – International Standard for Social Responsibility

Citation: (n.a.), 2018, "Discovering ISO 26000," Discovering ISO 26000 – Guidance on Social Responsibility.

Document Selection Substantiation

- **Global Influence:** Developed by the International Organization for Standardization, ISO 26000 provides guidance on social responsibility for all types of organizations and across all aspects of an organization's operations.
- **Regulatory Relevance:** While not a certifiable standard, rather a comprehensive guide to integrating social responsibility, it influences national standards and corporate policies. It is compliant with all applicable laws.
- **Practical Applicability:** Offers guidance on integrating social responsibility into organizational strategies, covering issues like

community involvement and labor practices. ISO 26000 helps businesses align their actions with societal expectations, promoting long-term sustainability and fostering positive relationships with stakeholders. It is designed to work in all organizations, in all regions and cultures.

- **Current Relevance:** Serves as a foundational document for organizations aiming to operate in a socially responsible manner, building a foundation for ethical and responsible business practices, contributing to the well-being of society and the planet. ISO 26000 is a reference document that provides guidance for organizations.

LEARNING OBJECTIVE

ISO 26000 is a globally recognized standard providing guidance on how businesses and organizations can operate in a socially responsible manner. Unlike other ISO standards, it is not intended for certification, but it offers a framework to help organizations integrate social responsibility (SR) into their operations and strategies. ISO 26000 emphasizes the importance of sustainability, ethical behavior, and stakeholder engagement in creating a positive impact on society and the environment. It covers seven core subjects and provides practical guidelines for addressing social, environmental, and economic challenges while improving organizational transparency and accountability.

UN Guiding Principles on Business and Human Rights

Citation: (n.a.), 2011, "Guiding Principles on Business and Human Rights: Implementing the United Nations 'Protect, Respect and Remedy' Framework," https://www.ohchr.org/documents/publications/guidingprinciplesbusinesshr_en.pdf.

Document Selection Substantiation

- **Global Influence:** Established by the United Nations, and consisting of 31 principles designed to implement the UN's "Protect, Respect and Remedy' framework," which is the global standard for preventing and addressing the risk of adverse human rights impacts linked to business activity.
- **Regulatory Relevance:** Inform national action plans and regulations on business and human rights. The UNPGs outline the expectation for companies to respect human rights and for governments to create frameworks that promote accountability and justice.

- **Practical Applicability:** The UNPGs clarify the roles of states and businesses and provide a framework for businesses to prevent and address human rights abuses in their operations and supply chains.
- **Current Relevance:** Increasingly referenced in legal contexts and by stakeholders demanding corporate accountability. By aligning with these principles, businesses can contribute to a more just and sustainable global economy while respecting fundamental human rights.

LEARNING OBJECTIVE

The UN Guiding Principles on Business and Human Rights (UNGPs), adopted in 2011, provide a global standard for preventing and addressing the risk of human rights harm linked to business activities. The principles were endorsed by the UN Human Rights Council and aim to clarify the duties of both states and businesses in relation to human rights. The framework consists of three main pillars: Protect, Respect, and Remedy.

World Bank's Environmental and Social Framework (ESF)

Citation: (n.a.), 2017, "The World Bank Environmental and Social Framework," ESFFramework.pdf.

Document Selection Substantiation

- **Global Influence:** The ESF sets mandatory social and environmental standards for projects funded by the World Bank, influencing policies and guidelines across global development initiatives. It is a key tool to help ensure that development projects are implemented responsibly and sustainably.
- **Regulatory Relevance:** Many governments, development banks, and institutions align their social and environmental policies with the World Bank's ESF standards. It helps borrowers identify, assess, and manage environmental and social risks.
- **Practical Applicability:** The ESF ensures that projects account for key social issues, such as labor rights, displacement, gender equality, stakeholder engagement, and indigenous rights. The ESF helps promote long-term positive outcomes for communities and the environment.
- **Current Relevance:** With the rise of responsible investing and sustainable development, the ESF continues to shape how projects integrate social safeguards and human rights considerations. Projects financed by the World Bank are more likely to meet global standards for social responsibility, environmental stewardship, and human rights.

LEARNING OBJECTIVE

The World Bank's Environmental and Social Framework (ESF) is a set of guidelines that outlines the environmental and social risks and requirements for projects financed by the World Bank. It is designed to promote sustainable development while safeguarding people and the environment. The ESF aims to ensure that the World Bank's projects do not cause harm to communities, ecosystems, or human rights and that they contribute positively to sustainable growth. The framework establishes requirements for managing risks and impacts related to environmental and social factors and requires transparency and accountability from borrowing countries and implementing organizations. It covers a broad range of topics, including labor and working conditions, biodiversity conservation, and indigenous rights.

GOVERNANCE

IFRS Sustainability-related Risks and Opportunities and the Disclosure of Material Information.

Citation: ISSB, November 2024, "IFRS Sustainability-related Risks and Opportunities and the Disclosure of Material Information," International Sustainability Standards Board, www.ifrs.org/content/dam/ifrs/supporting-implementation/issb-standards/issb-materiality-education-material.pdf.

Document Selection Substantiation

- **Global Influence:** This document carries significant global influence because it is issued by the International Sustainability Standards Board (ISSB). The ISSB aims to create a global baseline for sustainability-related financial disclosures. Their standards are intended to be used internationally, providing a common language for companies and investors worldwide. The document explains and illustrates how an entity might apply some of the requirements in IFRS Sustainability Disclosure Standards.
- **Regulatory Relevance:** The document directly supports the implementation of IFRS S1 and IFRS S2, which are the inaugural standards issued by the ISSB. These standards are increasingly being considered and adopted by jurisdictions around the world as a basis for mandatory sustainability reporting requirements. The document explains requirements related to identifying and disclosing material information about sustainability-related risks and opportunities. Furthermore, it addresses the interaction of ISSB standards with law or regulation in Appendix 3.

- **Practical Applicability:** The document is designed to be practically applicable for entities preparing sustainability-related financial disclosures. It includes illustrative examples and explanations to aid stakeholders' understanding of the standards. Chapter 3 provides a step-by-step guide for identifying and disclosing material information, enhancing its practical use.
- **Current Relevance:** With the increasing focus on ESG factors and the growing demand for transparent sustainability reporting, this document is highly relevant. The ISSB standards it supports are relatively new (issued in June 2023), making guidance on their application timely and crucial for entities seeking to comply with emerging reporting requirements. The document's date of November 2024 also confirms its current relevance.

LEARNING OBJECTIVE

Participants will be able to understand and apply the concept of materiality, as defined by the ISSB Standards (IFRS S1 and IFRS S2), to identify and disclose relevant information about sustainability-related risks and opportunities that could reasonably be expected to affect an entity's prospects, enabling informed decision-making by primary users of financial reports.

FC ESG Guidebook

Citation: Delange, N., et al., 2021, "IFC ESG Guidebook," International Finance Corporation, www.ifc.org/content/dam/ifc/doc/mgrt/ifc-esg-guidebook.pdf.

Document Selection Substantiation

- **Global Influence:** The International Finance Corporation (IFC), a member of the World Bank Group, is a globally recognized institution focused on private sector development in emerging markets. This guidebook reflects IFC's decades of experience in ESG and its role in shaping ESG concepts and standards. The document promotes international good practices in corporate governance, environmental and social management, and disclosure, contributing to a global understanding of ESG.
- **Regulatory Relevance:** The guidebook addresses the growing importance of ESG for regulatory compliance. It highlights the need for companies to manage ESG risks and build resilience. Although it does not directly create regulations, it supports the adoption of ESG

practices that align with evolving regulatory landscapes, particularly in emerging markets.

- **Practical Applicability:** The document is designed as a practical resource for companies in emerging markets looking to integrate ESG into their management and disclosure practices. It provides detailed guidance, good practices, and examples in the areas of governance, management, and disclosure. It includes frameworks, metrics, and indicators to facilitate ESG integration, making it directly applicable to companies seeking to improve their ESG performance.
- **Current Relevance:** The guidebook is relevant because it addresses the increasing focus on ESG issues driven by climate change, the COVID-19 pandemic, and the growing recognition of the importance of sustainable business practices. The document emphasizes the link between ESG practices and financial returns, which is a key consideration for investors and companies. The guide mentions various stakeholders collaborating to streamline and align standards, which is still relevant to current ESG development.

LEARNING OBJECTIVE

The test taker will be able to understand the core principles of ESG (Environmental, Social, and Governance), its relevance in emerging markets, and how to integrate ESG considerations into corporate governance, management practices, and disclosure strategies to enhance long-term value creation and resilience, as advocated by the International Finance Corporation (IFC).

Guide for Board Members: IFRS S1 and S2 Sustainability Standards

Citation: Instituto Brasileiro de Governança Corporativa, "Guide for Board Members: IFRS S1 and S2 Sustainability Standards," Instituto Brasileiro de Governança Corporativa, Chapter Zero Brazil, 55p.; PDF; 2.70 MB, guia_conselheiros_ING_P3.pdf.

Document Selection Substantiation

Global Influence

- **IFRS Foundation and ISSB:** The document addresses IFRS S1 and S2 standards developed by the International Sustainability Standards Board (ISSB) under the IFRS Foundation. The IFRS Foundation is

a globally recognized organization whose standards are increasingly adopted or used as a benchmark worldwide.
- **Climate Governance Initiative:** The guide is linked to the World Economic Forum's Climate Governance Initiative (CGI) through Chapter Zero Brazil. This connection provides access to a global network focused on climate governance.
- **IOSCO Endorsement:** The document mentions that the standards are endorsed by the International Organization of Securities Commissions (IOSCO), further amplifying their global influence among regulators and market participants.

Regulatory Relevance

- **Brazilian Adoption (CVM Resolution 193/2023):** The guide highlights that Brazil has adopted IFRS S1 and S2 through CVM Resolution 193/2023, making it directly relevant for Brazilian companies.
- **Future Regulations:** The document also anticipates that IFRS S1 and S2 are likely the first in a series of regulations that could cover issues related to biodiversity and human rights, signaling the increasing importance of sustainability reporting in the regulatory landscape.
- **Global Trend:** The guide notes that CVM's adoption of IFRS standards elevates Brazil to a pioneering position in sustainability reporting rules, following international standards. This underscores the regulatory relevance of these standards as other countries consider similar adoption. The document emphasizes mandatory assurance requirements (e.g., external audit obligations) for sustainability reports, ensuring alignment with CVM regulations.

Practical Applicability

- The guide is specifically designed for board members and governance agents, providing them with a tool to understand and implement IFRS S1 and S2. **The document provides a hands-on resource for board members,** helping them navigate IFRS S1 and S2 adoption.
- The guide outlines key objectives to help board members:
  - Monitor and validate the preparation of sustainability-related financial and non-financial reports.
  - Understand their fiduciary responsibility over non-financial information.
  - Develop strategic climate transition plans.
  - Understand broader ESG impacts.
  - Voluntarily adopt the standards in preparation for mandatory adoption.
  - The document includes summaries of the content of IFRS S1 and S2, making the standards more accessible.

- The guide provides structured implementation steps, including:
  - Establishing cross-functional governance committees.
  - Defining material sustainability risks and opportunities.
  - Integrating sustainability data with financial reporting.
  - Strengthening internal controls and risk management.

Current Relevance

- The guide addresses the urgent need for organizations to address climate, social, and environmental changes, which are highly relevant in the current global context.
- It emphasizes the importance of an integrated strategic approach to environmental, social, and climate issues, aligning with current expectations from investors, lenders, and regulators.
- The guide aims to standardize the disclosure of financial and non-financial information related to sustainability, making it clear, comparable, and consistent, which is crucial for a transparent and reliable global market.
- The document acknowledges that Brazil is at the forefront of the "green economy" and sustainable finance, and that climate change control, environmental preservation, and a sustainable agenda are cross-cutting themes in the capital market.
- The guide anticipates upcoming mandatory sustainability reporting deadlines (2026 onwards) and provides transition planning for organizations.
- It aligns with emerging sustainability assurance frameworks such as IAASB's ISSA 5000, which is expected to be finalized by the end of 2024.

Relevance to Organizations Outside of Brazil

- The document is highly relevant to organizations outside of Brazil attempting to implement sustainability disclosure requirements for the following reasons:
  - The core content focuses on IFRS S1 and S2, which are international standards. Organizations in countries that have adopted or are considering adopting these standards would find the guide directly applicable.
  - Even if a country hasn't fully adopted IFRS S1 and S2, the guide offers insights into best practices for sustainability reporting and governance. It can serve as a benchmark for organizations looking to improve their ESG disclosures.

- Investors are increasingly demanding standardized sustainability information. Understanding and implementing frameworks like IFRS S1 and S2 can help organizations attract international investment.
- The guide provides a structured approach to implementing these standards, including governance advancements, strategic implementation plans, and materiality assessments. This framework can be adapted to various organizational contexts.
- As sustainability reporting becomes more widespread and mandatory, organizations can use this guide to proactively prepare for future regulatory changes in their respective countries.
- For organizations outside of Brazil, the core principles in this document apply universally, as many regulators worldwide are aligning sustainability reporting with IFRS S1 and S2.

LEARNING OBJECTIVE

After reading this guide, board members and governance professionals will be better prepared to understand, assess, and implement IFRS S1 and S2 sustainability disclosure standards within their organizations. Further insight into developing the ability to integrate sustainability-related financial disclosures with corporate governance frameworks, ensuring compliance with regulatory requirements while identifying material Environmental, Social, and Governance (ESG) risks and opportunities that impact business strategy, financial stability, and investor decision-making, is also anticipated.

The Enhancement and Standardization of Climate-Related Disclosures for Investors

Citation: SEC, May 28, 2024, "The Enhancement and Standardization of Climate-Related Disclosures for Investors," Securities and Exchange Commission, www.sec.gov/files/rules/final/2024/33-11275.pdf.

Document Selection Substantiation

Global Influence

- While primarily focused on U.S. markets, the SEC's actions often have global repercussions due to the size and influence of the U.S. economy and capital markets. Many international companies are listed on U.S. exchanges, and U.S. regulations can influence global standards.

- The specific rules outlined (e.g., around GHG emissions, scenario analysis, etc.) may serve as a model for other countries developing their climate disclosure regulations.
- This SEC rule significantly influences global financial markets because the U.S. Securities and Exchange Commission (SEC) oversees companies that are key players in the global economy.
- The rule aligns partially with international ESG reporting frameworks, including:
  - TCFD (Task Force on Climate-Related Financial Disclosures).
  - GHG Protocol (Greenhouse Gas Accounting Standards).
  - ISSB (International Sustainability Standards Board) IFRS S2.

Regulatory Relevance
The document:

- Details a U.S. federal regulation, effective May 28, 2024, and applies to public companies filing under the Securities Act of 1933 and the Exchange Act of 1934. Details final rules that require registrants to provide specific climate-related information in their registration statements and annual reports. This is a significant development in mandatory climate disclosure.
- Addresses the SEC's authority to adopt these disclosure rules, reinforcing its regulatory power in this area.
- Outlines compliance dates and phase-in periods, making it crucial for companies subject to SEC regulations to understand and adhere to these rules.
- Cites specific CFR (Code of Federal Regulations) references, solidifying the legal and regulatory nature of the requirements.
- Highlights key reporting requirements that include:
  - Disclosure of severe weather event-related financial impacts.
  - Board oversight of climate risks.
  - Climate-related targets, transition plans, and risk management processes.
  - Scope 1 and Scope 2 GHG emissions disclosures for large registrants.

Practical Applicability
The document:

- Provides detailed guidance on what information must be disclosed (climate-related risks, impacts on strategy, governance, risk management, targets/goals, GHG emissions) and how it should be presented.

- Offers definitions of key terms like "climate-related risks" and "climate-related opportunities," providing clarity for companies assessing and reporting their climate impacts.
- Outlines guidance on the details of reporting GHG emissions, a focal point of climate-related accounting standards.
- Includes rules around attestation of GHG emissions disclosures, indicating a need for third-party verification in some cases.

Current Relevance

- The summary explicitly states that the rules are designed to enhance and standardize climate-related disclosures for investors, reflecting the increasing demand for this information in investment decisions.
- The document emphasizes disclosure of climate-related risks that have materially impacted or are reasonably likely to have a material impact on a company's business, reflecting the growing recognition of climate risk as a financial risk.
- The document notes that the final rules are effective on May 28, 2024, indicating current relevancy.
- The document provides information on transitioning to the final rules.
- The rule aligns with California's Climate Disclosure Laws (SB 253 and SB 261) and EU CSRD (Corporate Sustainability Reporting Directive) compliance requirements.
- Companies need to prepare now for the 2026 compliance deadline for reasonable assurance on emissions data.

Relevance to Organizations Outside of the United States

This document could be relevant to organizations outside of the United States attempting to implement sustainability disclosure requirements, specifically in the following ways:

- The SEC rules can serve as a benchmark for other countries or regions developing their climate disclosure standards. Policymakers and regulators can analyze the SEC's approach to scope, materiality, and attestation to inform their frameworks.
- Multinational corporations headquartered outside the U.S. but listed on U.S. exchanges will be directly subject to these rules.
- Even if not directly regulated, companies that want to attract investment from U.S. investors may choose to align their disclosures with the SEC rules to meet investor expectations.
- The SEC's extensive rulemaking process (including comments received and the SEC's responses) provides valuable insights into the challenges and considerations involved in developing mandatory climate disclosures.

- Organizations can use this information to learn about different climate disclosure standards and frameworks, enabling them to become more proactive in implementing various climate disclosure-related projects.
- The SEC's climate risk disclosure rules underscore the trend toward greater transparency and accountability regarding climate-related risks. Companies worldwide are likely to face growing pressure to disclose similar information, regardless of whether they are directly subject to the SEC rules.

LEARNING OBJECTIVE

Enable test takers to fully comprehend the SEC's final rules on climate-related disclosures for investors, including the scope of required information, presentation formats, attestation requirements, and compliance timelines, to ensure accurate and standardized reporting under U.S. securities laws.

OECD Guidelines for Multinational Enterprises on Responsible Business Conduct

Citation: OECD, June 8, 2023, "OECD Guidelines for Multinational Enterprises on Responsible Business Conduct," Organisation for Economic Co-operation and Development, OECD Publishing, Paris, www.oecd.org/content/dam/oecd/en/publications/reports/2023/06/oecd-guidelines-for-multinational-enterprises-on-responsible-business-conduct_a0b49990/81f92357-en.pdf.

Document Selection Substantiation

Global Influence

- The Organisation for Economic Co-operation and Development (OECD) is an intergovernmental organization with a significant global presence. Its recommendations carry weight and influence policy and business practices in many countries.
- The Guidelines are adhered to by all OECD member countries plus many non-member countries (Argentina, Brazil, Bulgaria, Croatia, Egypt, Jordan, Kazakhstan, Morocco, Peru, Romania, Tunisia, Ukraine, and Uruguay), extending their influence beyond the OECD membership.
- The Guidelines are developed with input from Business at OECD, the Trade Union Advisory Committee to the OECD, and OECD Watch, representing businesses, workers, and civil society.

- The Guidelines address a wide range of responsible business conduct issues, covering human rights, labor rights, the environment, anti-corruption, consumer interests, and more.
- The Guidelines align with global frameworks such as the United Nations Sustainable Development Goals (SDGs), the United Nations Guiding Principles on Business and Human Rights (UNGPs), and ILO labor standards.

Regulatory Relevance

- The Guidelines are supported by National Contact Points (NCPs) established by governments to promote their effectiveness. These NCPs provide a mechanism for addressing alleged breaches of the Guidelines.
- The Guidelines are recommendations from governments to multinational enterprises. While not directly legally binding in most cases, they reflect governmental expectations.
- The Guidelines often serve as a basis for national legislation or regulations related to responsible business conduct.
- The Guidelines align with and support various international agreements and standards, including the EU Corporate Sustainability Due Diligence Directive (CSDDD), German Supply Chain Act, and France's Duty of Vigilance Law. Use these guidelines as a reference.

Practical Applicability

- The Guidelines recommend that enterprises undertake risk-based due diligence to identify, prevent, mitigate, and account for how they address actual and potential adverse impacts.
- The document contains specific recommendations for enterprises in areas like human rights, labor relations, environmental protection, and combating corruption.
- The Guidelines include updated recommendations on the disclosure of responsible business conduct information.
- The guidance applies to multinational enterprises of all sizes operating in various sectors.

Current Relevance

- The fact that the Guidelines were updated in 2023 demonstrates their ongoing relevance and responsiveness to evolving societal challenges and business practices.
- The 2023 updates address pressing issues such as climate change, biodiversity, technology, and protection for whistleblowers.
- The Guidelines aim to enhance the business contribution to sustainable development and address adverse impacts on people, planet, and society.

- With increasing scrutiny of corporate behavior, the Guidelines offer a framework for companies to demonstrate their commitment to responsible business conduct.
- The Guidelines are particularly relevant as governments and investors demand greater transparency in corporate sustainability efforts, impacting investment decisions and risk management.

LEARNING OBJECTIVE

Upon completing the review of the OECD Guidelines for Multinational Enterprises on Responsible Business Conduct, test takers will be able to understand core principles, specific recommendations, and due diligence expectations across key areas such as human rights, environmental impact, labor, and anti-corruption. Test takers will gain insights into the implementation mechanisms, including the role of National Contact Points (NCPs), to promote responsible business practices and align with international sustainability frameworks.

Appendix 3
Glossary of ESG Terms and Definitions

Accounting Policies
The specific principles, bases, conventions, rules, and practices applied by an entity in preparing and presenting financial statements.

Adaptive Capacity
The ability of systems, institutions, humans, and other organisms to adjust to potential damage, to take advantage of opportunities, or to respond to consequences.

Afforestation
Planting of new forests on lands that historically have not contained forests.[1]

Agriculture, Forestry, and Other Land Use (AFOLU)
In the context of national greenhouse gas (GHG) inventories under the United Nations Framework Convention on Climate Change (UNFCCC), AFOLU is the sum of the GHG inventory sectors Agriculture and Land Use, Land-Use Change and Forestry (LULUCF).

Agroecology
The science and practice of applying ecological concepts, principles, and knowledge (i.e., the interactions of, and explanations for, the diversity, abundance, and activities of organisms) to the study, design, and management of sustainable agroecosystems. It includes the roles of human beings as a central organism in agroecology by way of social and economic processes in farming systems. Agroecology examines the roles and interactions among all relevant biophysical, technical, and socio-economic components of farming systems and their surrounding landscapes.

Albedo
The proportion of sunlight (solar radiation) reflected by a surface or object, often expressed as a percentage. Clouds, snow, and ice usually have high albedo; soil surfaces cover the albedo range from high to low; vegetation in the dry season and/or in arid zones can have high albedo, whereas photosynthetically active vegetation and the ocean have low

albedo. The Earth's planetary albedo changes mainly through changes in cloudiness and of snow, ice, leaf area, and land cover.

Alluvion

Torrential water flows dragging large amounts of solid material (pebbles, stones, and rock blocks) common in dry regions riverbeds, which are produced by heavy rain.

Anthropogenic Emissions

Emissions of greenhouse gases (GHGs), precursors of GHGs, and aerosols caused by human activities. These activities include the burning of fossil fuels, deforestation, land use and land-use changes (LULUC), livestock production, fertilization, waste management, and industrial processes. See also Anthropogenic and Anthropogenic removals.

Annual or Integrated Reports

Refer to reports that describe companies' activities for the preceding year (annual reports) or the broader range of measures that contribute to companies' long-term value and the role they play in society (integrated reports).

Avoid, Shift, Improve (ASI)

Reducing greenhouse gas emissions by avoiding the use of an emissions-producing service entirely, shifting to the lowest-emission mode of providing the service, and/or improving the technologies and systems for providing the service in ways that reduce emissions.[2]

Biomass

Organic material excluding the material that is fossilized or embedded in geological formations. Biomass may refer to the mass of organic matter in a specific area.[3]

Black Carbon (BC)

A relatively pure form of carbon, also known as soot, arising from the incomplete combustion of fossil fuels, biofuel, and biomass. It only stays in the atmosphere for days or weeks. BC is a climate forcing agent with a strong warming effect, both in the atmosphere and when deposited on snow or ice.[4]

Carbon Capture and Sequestration/Storage (CCS)

A technology that can capture carbon dioxide (CO_2) emissions produced from the use of fossil fuels in electricity generation and industrial processes and store the CO_2 deep underground, preventing the CO_2 from entering the atmosphere.[5]

Carbon Capture, Utilization, and Storage (CCS/CCUS)

CCUS technology is a new development trend of CCS technology. CCUS, in comparison to CCS, can recycle CO_2 resources for more effective carbon emission reduction.[6]

Carbon Footprint

Measure of the exclusive total amount of emissions of carbon dioxide (CO_2) that is directly and indirectly caused by an activity or is accumulated over the lifecycle stages of a product.[7]

Carbon Footprinting

Is the calculation of the total greenhouse gas emissions caused by an individual, event, organization, service, or product, expressed as carbon dioxide equivalent.

Carbon Intensity

Relates to a company's physical carbon performance and describes the extent to which its business activities are based on carbon usage for a defined scope and fiscal year.

Carbon Neutrality

Condition in which anthropogenic carbon dioxide (CO_2) emissions associated with a subject are balanced by anthropogenic CO_2 removals. The subject can be an entity such as a country, an organization, a district or a commodity, or an activity such as a service and an event. Carbon neutrality is often assessed over the lifecycle including indirect ("scope 3") emissions but can also be limited to the emissions and removals, over a specified period, for which the subject has direct control, as determined by the relevant scheme.[8]

Carbon Price

The price for avoided or released carbon dioxide (CO_2) or CO_2-equivalent emissions. This may refer to the rate of a carbon tax or the price of emission permits. In many models that are used to assess the economic costs of mitigation, carbon prices are used as a proxy to represent the level of effort in mitigation policies.[9]

Climate

Climate in a narrow sense is usually defined as the average weather, or more rigorously, as the statistical description in terms of the mean and variability of relevant quantities over a period of time ranging from months to thousands or millions of years. The classical period for averaging these variables is 30 years, as defined by the World Meteorological Organization (WMO). The relevant quantities are most often surface variables such as temperature, precipitation, and wind. Climate in a wider sense is the state, including a statistical description, of the climate system.[10]

Climate Change

A change in the state of the climate that can be identified (e.g., by using statistical tests) by changes in the mean and or the variability of its properties and that persists for an extended period, typically decades or longer. Climate change may be due to natural internal processes or external forcings such as modulations of the solar cycles, volcanic eruptions, and persistent anthropogenic changes in the composition of the atmosphere or in land use. Note that the United Nations Framework Convention on Climate Change (UNFCCC), in its Article 1, defines climate change as: "a change of climate which is attributed directly or indirectly to human activity that alters the composition of the global atmosphere and which is in addition to natural climate variability observed over comparable time periods." The UNFCCC thus makes a distinction between climate change attributable to human activities altering the atmospheric composition and climate variability attributable to natural causes.[11]

Climate Change Commitment

The unavoidable future climate change resulting from inertia in the geophysical and socio-economic systems. Different types of climate change commitment are discussed in the literature. Climate change commitment is usually quantified in terms of the further change in temperature, but it includes other future changes, for example in the hydrological cycle, in extreme weather events, in extreme climate events, and in sea level.[12]

Climate Model

A qualitative or quantitative representation of the climate system based on the physical, chemical, and biological properties of its components, their interactions and feedback processes and accounting for some of its known properties. The climate system can be represented by models of varying complexity; that is, for any one component or combination of components a spectrum or hierarchy of models can be identified, differing in such aspects as the number of spatial dimensions, the extent to which physical, chemical, or biological processes are explicitly represented, or the level at which empirical parametrizations are involved. There is an evolution toward more complex models with interactive chemistry and biology. Climate models are applied as a research tool to study and simulate the climate and for operational purposes, including monthly, seasonal, and interannual climate predictions.[13]

Climate-Related Opportunity

Refers to the potential positive impacts related to climate change on an organization. Efforts to mitigate and adapt to climate change can produce opportunities for organizations, such as through resource efficiency and cost savings, the adoption and utilization of low-emission energy sources, the development of new products and services, and building resilience along the supply chain. Climate-related opportunities will vary depending on the region, market, and industry in which an organization operates.

Climate-Related Risk

Refers to the potential negative impacts of climate change on an organization. Physical risks emanating from climate change can be event-driven (acute) such as increased severity of extreme weather events (e.g., cyclones, droughts, floods, and fires). They can also relate to longer-term shifts (chronic) in precipitation and temperature and increased variability in weather patterns (e.g., sea level rise). Climate-related risks can also be associated with the transition to a lower-carbon global economy, the most common of which relate to policy and legal actions, technology changes, market responses, and reputational considerations.

Climate Variability

Deviations of climate variables from a given mean state (including the occurrence of extremes, etc.) at all spatial and temporal scales beyond that of individual weather events. Variability may be intrinsic, due to fluctuations of processes internal to the climate system (internal variability), or extrinsic, due to variations in natural or anthropogenic external forcing (forced variability).

CMIP5 Protocol

Is a set of 35 climate model experiments designed to be useful in (1) assessing the mechanisms responsible for model differences in poorly understood feedbacks associated with the carbon cycle and with clouds, (2) examining climate "predictability" and exploring the ability of models to predict climate on decadal time scales, and, more generally, (3) determining why similarly forced models produce a range of responses.[14]

CO_2 equivalent (CO_2-eq) Emission

The amount of carbon dioxide (CO_2) emission that would have an equivalent effect on a specified key measure of climate change, over a specified time horizon, as an emitted amount of another greenhouse gas (GHG) or a mixture of other GHGs. For a mix of GHGs, it is obtained by summing the CO_2-equivalent emissions of each gas. There are various ways and time horizons to compute such equivalent emissions (see greenhouse gas emission metric). CO_2-equivalent emissions are commonly used to compare emissions of different GHGs but should not be taken to imply that these emissions have an equivalent effect across all key measures of climate change.[15]

Conference of the Parties (COP)

The supreme body of UN conventions, such as the United Nations Framework Convention on Climate Change (UNFCCC), comprising parties with a right to vote that have ratified or acceded to the convention.

Cumulative Emissions

The total amount of emissions released over a specified period of time.

Decarbonization

Human actions to reduce carbon dioxide emissions from human activities.

Decoupling

In relation to climate change, is where economic growth is no longer strongly associated with another relevant indicator such as greenhouse gas emissions.

Desertification

Land degradation in arid, semi-arid, and dry sub-humid areas resulting from many factors, including climatic variations and human activities.

Designated Index

Any of the Standard & Poor's 500 Composite Stock Price Index, the Standard & Poor's 500 Total Return Index, the Nasdaq-100 Index or any other Index, with the goal of matching the performance of a securities portfolio against that benchmark (designated index), as closely as possible.

Direct Air Capture (DAC)

Chemical process by which a pure carbon dioxide (CO_2) stream is produced by capturing CO_2 from the ambient air.

Disaster

A serious disruption of the functioning of a community or a society at any scale due to hazardous events interacting with conditions of exposure, vulnerability, and capacity, leading to one or more of the following: human, material, economic and environmental losses, and impacts.[16]

Double Materiality

Calls on organizations to consider materiality from two viewpoints: financial materiality and materiality to the market, the environment, and people.

Double materiality recognizes that an organization is responsible for managing its own financial risks by looking inward. But it also looks at the outward impacts of its decisions and operations on people and the environment.

By applying the concept of double materiality, organizations can identify both the financial and non-financial impacts of their operations to help shape a more holistic ESG strategy.[17]

Energy Efficiency

The ratio of output or useful energy or energy services or other useful physical outputs obtained from a system, conversion process, transmission, or storage activity to the input of energy (measured as kWh kWh^{-1}, tons kWh^{-1} or any other physical measure of useful output like ton-km transported). Energy efficiency is often described by energy intensity.

Energy Security
The goal of a given country, or the global community as a whole, to maintain an adequate, stable, and predictable energy supply. Measures encompass safeguarding the sufficiency of energy resources to meet national energy demand at competitive and stable prices and the resilience of the energy supply; enabling development and deployment of technologies; building sufficient infrastructure to generate, store, and transmit energy supplies and ensuring enforceable contracts of delivery.[18]

Enhanced Weathering
A proposed method to increase the natural rate of removal of carbon dioxide (CO_2) from the atmosphere using silicate and carbonate rocks. The active surface area of these minerals is increased by grinding before they are actively added to soil, beaches, or the open ocean.

Extreme Weather Event
An event that is rare at a particular place and time of year. Definitions of "rare" vary, but an extreme weather event would normally be as rare as, or rarer than, the 10th or 90th percentile of a probability density function estimated from observations. By definition, the characteristics of what is called extreme weather may vary from place to place in an absolute sense.[19]

Emissions Scenario
A plausible future pathway of man-made emissions (e.g., greenhouse gases and other pollutants) that can affect climate. These pathways are based on a coherent and internally consistent set of assumptions about determining factors (such as demographic and socioeconomic development, technological change) and their key relationships.[20]

Energy Transition
A shift from a system currently dominated by mainly fossil-fuel-based energy toward a system using a majority of low-emission and renewable energy sources and maximizing opportunities for increased energy efficiency and better management of energy demand.[21]

Energy and Natural Resources (ENR)
The ENR sector includes oil and gas, power and utilities, mining, and forestry industries.

Forecasting
Is based on past and present data and analysis of trends. Often it takes the form of predicting a single, most probable trend for and into the future.[22]

Fossil Fuels
Carbon-based fuels from fossil hydrocarbon deposits, including coal, oil, and natural gas.

Fugitive Emissions (Oil and Natural Gas Systems)
The release of greenhouse gases that occurs during the exploration, processing, and delivery of fossil fuels to the point of final use. This excludes greenhouse gas emissions from fuel combustion for the production of useful heat or power. It encompasses venting, flaring, and leaks.[23]

General Circulation Models (GCM)
These are numerical models representing physical processes in the atmosphere, ocean, cryosphere, and land surface.[24]

Global Warming
Refers to the increase in global surface temperature relative to a baseline reference period, averaging over a period sufficient to remove interannual variations (e.g., 20 or 30 years). A common choice for the baseline is 1850–1900 (the earliest period of reliable observations with sufficient geographic coverage), with more modern baselines used depending upon the application.[25]

Governance
Refers to "the system by which an organization is directed and controlled in the interests of shareholders and other stakeholders."[26]
Governance involves a set of relationships between an organization's management, its board, its shareholders, and other stakeholders. Governance provides the structure and processes through which the objectives of the organization are set, progress against performance is monitored, and results are evaluated.[27]

Green Climate Fund (GCF)
The GCF was established by the 16th Session of the Conference of the Parties (COP) in 2010 as an operating entity of the financial mechanism of the United Nations Framework Convention on Climate Change (UNFCCC), in accordance with Article 11 of the Convention, to support projects, programs and policies and other activities in developing country Parties. The Fund is governed by a Board and will receive guidance of the COP.[28]

Green Deal
Established by the European Commission is a set of policy initiatives and green investments to make Europe carbon-neutral by 2050.

Green Deposit
A fixed or a floating-term deposit for investors looking to invest their surplus cash balance in environmentally friendly and sustainable projects.

This type of investment is associated with assets and projects consistent with delivering a low carbon and climate-resilient economy.

Green Guarantees and Letters of Credit

Are trade finance instruments which support, guarantee and/or finance an underlying project having a clear positive contribution to the environment.

Greenhouse Gases (GHGs)

Gaseous constituents of the atmosphere, both natural and anthropogenic, that absorb and emit radiation at specific wavelengths within the spectrum of radiation emitted by the Earth's surface, by the atmosphere itself, and by clouds. This property causes the greenhouse effect. Water vapor (H_2O), carbon dioxide (CO_2), nitrous oxide (N_2O), methane (CH_4), and ozone (O_3) are the primary GHGs in the Earth's atmosphere. Humanmade GHGs include sulfur hexafluoride (SF_6), hydrofluorocarbons (HFCs), chlorofluorocarbons (CFCs), and perfluorocarbons (PFCs); several of these are also O_3-depleting (and are regulated under the Montreal Protocol).[29]

Greenhouse Gas (GHG) Emissions Inventory

A list of emission sources and the associated emissions quantified using standardized methods. Organizations develop GHG inventories for a variety of reasons, including: Managing GHG risks and identifying reduction opportunities. Participating in voluntary or mandatory GHG programs.

GHG emissions inventory – incorporating scope 1, scope 2, and scope 3 emissions – enables companies to understand their full emissions impact across the value chain and focus efforts where they can have the greatest impact.[30]

Greenhouse Gas (GHG) Emissions Scope Levels[31]

- Scope 1 refers to all direct GHG emissions.
- Scope 2 refers to indirect GHG emissions from the consumption of purchased electricity, heat, or steam.
- Scope 3 refers to other indirect emissions not covered in Scope 2 that occur in the value chain of the reporting company, including both upstream and downstream emissions. Scope 3 emissions could include: the extraction and production of purchased materials and fuels, transport-related activities in vehicles not owned or controlled by the reporting entity, electricity-related activities (e.g., transmission and distribution losses), outsourced activities, and waste disposal.[32,33]

Greenhouse Gas (GHG) Protocol Methodology

Is the most widely recognized and used international standard for calculating GHG emissions.

Greenwashing

Is a misrepresentation of a product, service, or investment, making something appear to be more sustainable than it actually is. Greenwashing includes eggs labeled "farm fresh," clothing marketed as "eco-friendly" and investments labeled "green" when those claims are either exaggerated or cannot be substantiated.[34]

Human Rights

Rights that are inherent to all human beings, universal, inalienable, and indivisible, typically expressed and guaranteed by law. They include the right to life, economic, social, and cultural rights, and the right to development and self-determination.[35]

Index Fund

A type of mutual fund or exchange-traded fund (ETF) that holds all (or a representative sample) of the securities in a specific index (e.g., the Standard & Poor's 500 Index, Dow Jones Industrial Average, Russell 2000 Index, FTSE 100 Index, Nasdaq 100 Index, Wilshire 5000 Total Market Index, MSCI EAFE Index, Bloomberg U.S. Aggregate Bond Index, etc.), with the goal of matching the performance of that benchmark as closely as possible.

Indigenous Knowledge

The understandings, skills, and philosophies developed by societies with long histories of interaction with their natural surroundings. For many indigenous peoples, indigenous knowledge informs decision-making about fundamental aspects of life, from day-to-day activities to longer-term actions. This knowledge is integral to cultural complexes, which also encompass language, systems of classification, resource use practices, social interactions, values, ritual, and spirituality. These distinctive ways of knowing are important facets of the world's cultural diversity.[36]

Insurance-Based Investment Product (IBIP)

An insurance product that is made available to a professional investor and which offers a maturity or surrender value that is wholly or partially exposed, directly or indirectly, to market fluctuations.[37]

Integrated Assessment Models (IAM)

These models attempt to integrate knowledge from two or more domains of expertise or academic disciplines. They are constructed to address climate change by tracking emissions, the concentration of greenhouse gases in the atmosphere as well as other carbon sinks, temperature and other climate impacts arising from increased concentrations of greenhouse gases in the atmosphere, and damages resulting from those climate impacts. Emissions follow from economic behavior, and policy scenarios can be hypothesized to affect emissions along several dimensions.[38]

Internal Carbon Price

Refers to a monetary value on greenhouse gas (GHG) emissions that an organization uses internally to guide its decision-making process in relation to climate change impacts, risks, and opportunities.

International Energy Agency (IEA) 2DS

The 2°C Scenario 2DS describes an energy system consistent with an emissions trajectory that recent climate science research indicates would give an 80% chance of limiting average global temperature increase to 2°C.

2°C scenario asks the question, "if the world limits warming at or below 2°C, what are the pathways for achieving that goal?"

Implied Temperature Rise (ITR)

Refers to an estimate of a global temperature rise associated with the greenhouse gas emissions of a single entity (e.g., a company) or a selection of entities (e.g., those in each investment portfolio, fund, or investment strategy). Expressed as a numeric degree rating, ITR metrics incorporate current GHG emissions or other data and assumptions to estimate expected future emissions associated with the selected entity or entities. Then, the estimate is translated into a projected increase in global average temperature (in °C) above preindustrial levels that would occur if all companies in corresponding sectors had the same carbon intensity as the selected asset(s).

Kaya Identity

In this identity, global emissions are equal to the population size, multiplied by per capita output (gross world product), multiplied by the energy intensity of production, multiplied by the carbon intensity of energy.[39]

Land Use/Land Use Change/Forestry (LULUCF)

A greenhouse gas inventory sector that covers emissions and removals of greenhouse gases resulting from direct human-induced land use, land-use change, and forestry activities.[40]

Liquefaction

Soil liquefaction describes a phenomenon whereby a saturated soil substantially loses strength and stiffness in response to an applied stress, usually earthquake shaking or other sudden change in stress condition, causing it to behave like a liquid.

Mainstream Index

A single equity index for direct exposure to a country, region, or sector, such as the Eurostoxx-50, S&P-500, Nasdaq-100, Nikkei 225, and FTSE-100. Mainstream indices have established familiarity and usage over many decades. Mainstream indices maintain a strong position as the choice for most retail investors.

Maladaptive Actions (Maladaptation)

Actions that may lead to increased risk of adverse climate-related outcomes, including via increased greenhouse gas (GHG) emissions, increased vulnerability to climate change, or diminished welfare, now or in the future. Maladaptation is usually an unintended consequence.[41]

Material

Refers to information that is important, relevant, and/or critical to long-term value creation. The issue of materiality and what should be disclosed in annual reports varies according to regional regulations and expectations, particularly in the United States.

Materiality in the Context of ESG?

The concept of materiality guides organizations to focus on ESG issues that are relevant to them and will have a measurable impact on their business.

To determine materiality, an organization must first identify its risks and then assess the consequences of those vulnerabilities. Using a "risk matrix" approach, organizations can determine which ESG-related risks to prioritize based on their risk profile, and which of those consequences would have significant negative impacts on their organization.[42]

Methane (CH_4)

The greenhouse gas (GHG) methane is the major component of natural gas and is associated with all hydrocarbon fuels. Significant anthropogenic emissions also occur because of animal husbandry and paddy rice production. Methane is also produced naturally where organic matter decays under anaerobic conditions, such as in wetlands. Under future global warming, there is potential for increased methane emissions from thawing permafrost, wetlands, and sub-sea gas hydrates.[43]

Mitigation Measures

In climate policy, mitigation measures are technologies, processes, or practices that contribute to mitigation, for example, renewable energy technologies, waste minimization processes, and public transport commuting practices.

Nationally Determined Contribution (NDC)

Refers to the post-2020 actions that a country intends to take under the international climate agreement adopted in Paris.

Net Negative Greenhouse Gas Emissions

A situation of net negative greenhouse gas emissions is achieved when metric-weighted anthropogenic greenhouse gas (GHG) removals exceed metric-weighted anthropogenic GHG emissions. Where multiple GHG are involved, the quantification of net emissions depends on the metric chosen to compare emissions of different gases (such as global warming potential, global temperature change potential, and others, as well as the chosen time horizon).[44]

Net-Zero

Refers to achieving an equal balance between Greenhouse Gas (GHG) emissions produced and GHG emissions removed from the atmosphere.

Net-Zero CO_2 Emissions

A condition in which anthropogenic carbon dioxide (CO_2) emissions are balanced by anthropogenic CO_2 removals over a specified period.

Net-Zero Greenhouse Gas Emissions

A condition in which metric-weighted anthropogenic greenhouse gas (GHG) emissions are balanced by metric-weighted anthropogenic GHG removals over a specified period. The quantification of net-zero GHG emissions depends on the GHG emission metric chosen to compare emissions and removals of different gases, as well as the time horizon chosen for that metric.[45]

Nitrous Oxide (N_2O)

The main anthropogenic source of N_2O, a greenhouse gas (GHG), is agriculture (soil and animal manure management), but important contributions also come from sewage treatment, fossil fuel combustion, and chemical industrial processes. N_2O is also produced naturally from a wide variety of biological sources in soil and water, particularly microbial action in wet tropical forests.[46]

Ocean Alkalinization/Ocean Alkalinity Enhancement

A proposed carbon dioxide removal (CDR) method that involves deposition of alkaline minerals or their dissociation products at the ocean surface. This increases surface total alkalinity and may thus increase ocean carbon dioxide (CO_2) uptake and ameliorate surface ocean acidification.[47]

Oxygen Minimum Zone (OMZ)

The midwater layer (200–1000 m) in the open ocean in which oxygen saturation is the lowest in the ocean. The degree of oxygen depletion depends on the largely bacterial consumption of organic matter and the distribution of the OMZs is influenced by large-scale ocean circulation. In coastal oceans, OMZs extend to the shelves and may also affect benthic ecosystems.[48]

Ozone (O_3)

The triatomic form of oxygen and a gaseous atmospheric constituent. In the troposphere, O_3 is created both naturally and by photochemical reactions involving gases resulting from human activities (e.g., smog).

Tropospheric O_3 acts as a greenhouse gas (GHG). In the stratosphere, O_3 is created by the interaction between solar ultraviolet radiation and molecular oxygen (O_2). Stratospheric O_3 plays a dominant role in the stratospheric radiative balance. Its concentration is highest in the ozone layer.[49]

Pareto Optimum
A state in which no one's welfare can be increased without reducing someone else's welfare.

Paris Agreement
In 2015, Parties to the UNFCCC agreed in Paris to keep the global temperature rise this century well below 2°C above pre-industrial levels and to pursue efforts to limit the temperature increase even further to 1.5°C. The agreement requires all Parties to put forward "Nationally Determined Contributions" (NDCs). There will also be a global stocktaking every five years to assess the collective progress toward achieving the agreement and to inform about further individual actions by Parties.[50]

Precursors
Atmospheric compounds that are not greenhouse gases (GHGs) or aerosols, but that influence GHG or aerosol concentrations by taking part in physical or chemical processes regulating their production or destruction rates.

Principal Adverse Impacts (PAIs)

- Negative, material, or potentially material effects on sustainability factors that result from, worsen, or are directly related to investment choices or advice performed by a legal entity. Examples include GHG emissions and carbon footprint.
- Principal Adverse Impact indicators are essentially a set of mandatory indicators and metrics which aim to show financial market participants how certain investments pose sustainability risks.
- The principal adverse impact indicators require FMPs to provide extensive disclosures on various ESG-related matters including environmental and social indicators including human rights.[51]

Publicly Available 2°C Scenario
Refers to a 2°C scenario that is (1) used/referenced and issued by an independent body; (2) wherever possible, supported by publicly available datasets; (3) updated on a regular basis; and (4) linked to functional tools (e.g., visualizers, calculators, and mapping tools) that can be applied by organizations. 2°C scenarios that presently meet these criteria include IEA 2DS, IEA 450, Deep Decarbonization Pathways Project, and International Renewable Energy Agency.

Qualitative Assessment

Uses subjective judgment to analyze a company's value or prospects based on non-quantifiable information, such as management expertise, industry cycles, strength of research and development, and labor relations.

Quantitative Assessment

Use of a set of methods, principles, or rules for assessment, based on the use of statistical analyses of numerical data obtained from surveys, experiments, or administrative records where the meanings and proportionality of values are maintained inside and outside the context of the assessment. Inferences can be made and correlations between variables analyzed to understand more generalized phenomena.

Reducing Emissions from Deforestation and Forest Degradation (REDD+)

REDD+ refers to reducing emissions from deforestation; reducing emissions from forest degradation; conservation of forest carbon stocks; sustainable management of forests; and enhancement of forest carbon stocks.[52]

Renewable Energy (RE)

Any form of energy that is replenished by natural processes at a rate that equals or exceeds its rate of use.

Representative Concentration Pathways (RCPs)

Four independent pathways comprising sets of projections of radiative forcing that serve as inputs to climate modeling, pattern scaling, and atmospheric chemistry modeling. These are based on the forcing of greenhouse gases and other forcing agents.[53]

Risk

The possibility or likelihood that actual results (operational or financial) deviate from expected results in a manner that influences objectives at different levels (such as strategic, organization-wide, project, product, and process).

Risk is often characterized by reference to potential events and consequences, or a combination of these, and expressed in terms of a combination of the consequences of an event (including changes in circumstances) and the associated likelihood of occurrence. Uncertainty is the state, even partial, of deficiency of information related to, understanding or knowledge of, an event, its consequence, or likelihood.

Risk conceptually equals the probability or likelihood of hazardous events occurring multiplied by the company's exposure and vulnerability to the event.[54]

Risk Assessment

Consists of risks identification, risk analysis, and risk evaluation. The essential building blocks for comprehensively assessing risk (and establishing metrics) are hazards, exposure, vulnerability, risk, and impacts.

Risk Management

Refers to a set of processes that are carried out by a company or organization's board and management to support the achievement of its objectives by addressing its risks and managing the combined potential impact of those risks.

Risk Profile

Refers to each company's risk attitude, which may be described as risk-averse, risk-neutral, or risk-seeking, a risk tolerance, which looks at acceptable/unacceptable deviations from what is expected, and a risk appetite, which looks at how much risk one is willing to accept.

Scenario

A scenario describes a path of development leading to a particular outcome. Scenarios are not intended to represent a full description of the future, but rather to highlight central elements of a possible future and to draw attention to the key factors that will drive future developments. It is important to remember that scenarios are hypothetical constructs; they are not forecasts or predictions, nor are they sensitivity analyses.

A key feature of scenarios is that they should challenge conventional wisdom about the future. In a world of uncertainty, scenarios are intended to explore alternatives that may significantly alter the basis for "business-as-usual" assumptions.

Scenario analysis differs from techniques such as sensitivity analysis, forecasting, or value at risk (VaR).[55]

Scenario Analysis

The purpose of scenario analysis is to consider and better understand how a business might perform under different future states (i.e., its robustness). In the case of climate change, climate-related scenarios allow an organization to explore and develop an understanding of how the physical and transition risks and opportunities of climate change might impact the business over time. Scenario analysis, therefore, evaluates a range of potential outcomes by considering a variety of alternative plausible future states (scenarios) under a given set of assumptions and constraints.

In this context, robustness refers to the resilience or ability of an organization's business strategy to tolerate disruptions or adapt to changes or uncertainties in the business environment that might affect the organization's performance and to remain effective under all or most situations and conditions.[56]

Scenario – Emissions

A plausible representation of the future development of emissions of substances that are radiatively active (e.g., greenhouse gases or aerosols), plus human-induced land-cover changes that can be radiatively active via albedo changes, based on a coherent and internally consistent set of assumptions about driving forces (such as demographic and socio-economic development, technological change, energy, and land use) and their key relationships. Concentration scenarios, derived from emission scenarios, are often used as input to a climate model to compute climate projections.[57]

Sector

A segment of companies performing similar business activities in an economy. A sector generally refers to a large segment of the economy or grouping of business types, while "industry" is used to describe more specific groupings of companies within a sector.

Sensitivity Analysis

Is the process of recalculating outcomes under alternative assumptions to determine the impact of a particular variable.[58]

Social Inclusion

A process of improving the terms of participation in society, particularly for people who are disadvantaged, through enhancing opportunities, access to resources, and respect for rights.[59]

Socially Responsible Investing (SRI)

An investment strategy that highlights this one facet of ESG. SRI investors seek companies that promote ethical and socially conscious themes including diversity, inclusion, community-focus, social justice, and corporate ethics, in addition to fighting against racial, gender, and sexual discrimination.

Soil Erosion

Washing away of soil down the surface of hill slopes or mass movements due to storm water flow during intense rains or winds. This can cause in turn sedimentation in streams/rivers and areas at the toe of the hills.

Stranded Assets

In the build environment are properties that will be increasingly exposed to the risk of early economic obsolescence due to climate change because they will not meet (potential) future regulatory efficiency standards or market expectations. These buildings will become less marketable and may require costly refurbishment measures.

Stranding Risk

Comprises potential write-downs due to direct climate change impacts and devaluations related to the transition to a "low-carbon economy."

Strategy

Speaks to an organization's desired future state. An organization's strategy establishes a foundation against which it can monitor and measure its progress in reaching that desired state. Strategy formulation generally involves establishing the purpose and scope of the organization's activities and the nature of its businesses, considering the risks and opportunities it faces and the environment where it operates.

Structural Collapse

Damages or collapse of any type of structure for reasons such as excess weight in public places, bridges, etc. This event includes damage that, although not taking the structures to the point of collapse, does make them unusable. Damages in structures caused by natural phenomena are reported as an effect of these phenomena.

Sustainability

A dynamic process that guarantees the persistence of natural and human systems in an equitable manner.

Sustainable Development (SD)

Development that meets the needs of the present without compromising the ability of future generations to meet their own needs and balances social, economic, and environmental concerns.[60]

Sustainable Development Goals (SDGs)

The 17 global goals for development for all countries established by the United Nations through a participatory process and elaborated in the 2030 Agenda for Sustainable Development, including ending poverty and hunger; ensuring health and well-being, education, gender equality, clean water and energy, and decent work; building and ensuring resilient and sustainable infrastructure, cities, and consumption; reducing inequalities; protecting land and water ecosystems; promoting peace, justice, and partnerships; and taking urgent action on climate change.[61]

Sustainable Investment

An investment in an economic activity that contributes to:

- An environmental objective, as measured, for example, by key resource efficiency indicators on the use of energy, renewable energy, raw materials, water, and land, on the production of waste, and greenhouse gas emissions, or on its impact on biodiversity and the circular economy.
- A social objective, in particular an investment that contributes to tackling inequality or that fosters social cohesion, social integration, and labor relations.

An investment in human capital or economically or socially disadvantaged communities, provided that such investments do not significantly harm any of those objectives and that the investee companies follow good governance practices, with respect to sound management structures, employee relations, remuneration of staff, and tax compliance.[62]

Sustainability Index

An instrument that measures the responsibility of a certain company in social and environmental areas. The more they take these aspects into account as they develop their business, the higher the score they will obtain.

Sustainability Indices (Examples)

The Dow Jones Sustainability™ World Index – Comprises global sustainability leaders of the top 10% of the largest 2,500 companies in the S&P Global BMI based on long-term economic, environmental, and social criteria. A defined set of criteria is used to assess the economic, social, and environmental opportunities of the companies that the DJSI has listed.

The Food Sustainability Index (FSI) – Determines how food systems are performing across three pillars: food loss and waste, sustainable agriculture, and nutritional challenges. Its 38 indicators and 95 sub-indicators address societal, environmental, and economic themes in 78 countries.

Environmental Sustainability Index – A global sampling of more than 600 business leaders directly involved in their organization's environmental sustainability initiatives, measuring their perception of how well their organization has performed in achieving its goals over the past year and optimism for the year ahead.

Global Sustainability Index – A diversified, sector-neutral global benchmark of organizations that are selected based on sector ranking of environmental, social, and government factors.

Sustainability Report

A report that describes a company's or organization's impact on society, often addressing Environmental, Social, and Governance issues.

Sustainability Risk

An Environmental, Social, or Governance event or condition that, if it occurs, could cause an actual or a potential material negative impact on the value of the investment.

Technology Deployment

The act of bringing technology into effective application, involving a set of actors and activities to initiate, facilitate, and/or support its implementation.

Transition Plan

Refers to an aspect of an organization's overall business strategy that lays out a set of targets and actions supporting its transition toward a low-carbon economy, including actions such as reducing its Greenhouse Gas (GHG) emissions.

Transition Risks

Transition risks are business-related risks that follow societal and economic shifts toward a low-carbon and more climate-friendly future. The risks relate to the economic and societal transformation needed to reach net zero emissions. These risks can include policy and regulatory risks, technological risks, market risks, reputational risks, and legal risks. These risks are interconnected and often top of mind for investors as they attempt to navigate an increasingly aggressive low-carbon agenda that can create capital and operational consequences to their assets.[63]

United Nations Convention to Combat Desertification (UNCCD)

A legally binding international agreement linking environment and development to sustainable land management, established in 1994. The Convention's objective is "to combat desertification and mitigate the effects of drought in countries experiencing drought and/or desertification." The Convention specifically addresses the arid, semi-arid and dry sub-humid areas, known as the drylands, and has a particular focus on Africa. As of September 2020, the UNCCD had 197 Parties.

United Nations Framework Convention on Climate Change (UNFCCC)

The UNFCCC was adopted in May 1992 and opened for signature at the 1992 Earth Summit in Rio de Janeiro. It entered into force in March 1994 and, as of September 2020, had 197 Parties (196 States and the European Union). The Convention's ultimate objective is the "stabilization of greenhouse gas concentrations in the atmosphere at a level that would prevent dangerous anthropogenic interference with the climate system." The provisions of the Convention are pursued and implemented by two further treaties: the Kyoto Protocol and the Paris Agreement.[64]

Value at Risk

Measures the size of financial loss a given portfolio might experience within a given time horizon and for a particular probability. Climate VaR has a long-time horizon (many years) compared with the shorter time horizon of standard financial VaR.[65]

Vulnerability

The propensity or predisposition to be adversely affected. Vulnerability encompasses a variety of concepts and elements including sensitivity or susceptibility to harm and lack of capacity to cope and adapt.

Wind Energy

Kinetic energy from airflow arising from the uneven heating of the Earth's surface. The wind's kinetic energy is converted to mechanical shaft energy and electricity by a wind turbine, a rotating machine. A wind farm, wind project, wind park, or wind power plant is a group of wind turbines interconnected to a common utility system through a system of transformers, distribution lines, and (usually) one substation.[66]

Zero Emissions Commitment

The zero emissions commitment is an estimate of the subsequent global warming that would result after anthropogenic emissions are set to zero. It is determined by both inertia in physical climate system components (ocean, cryosphere, land surface) and carbon cycle inertia. In its widest sense, it refers to emissions of each climate forcer including greenhouse gases, aerosols, and their precursors. The climate response to this can be complex due to the different climate response time scale of each climate forcer. A specific sub-category of zero emissions commitment is the Zero CO_2 Emissions Commitment, which refers to the climate system response to CO_2 emissions after setting these to net zero. The CO_2-only definition is of specific use in estimating remaining carbon budgets.[67]

NOTES

1 IPCC, 2014, Annex II: Glossary [Mach, K.J., S. Planton, and C. von Stechow (eds.)]. In: *Climate Change 2014: Synthesis Report. Contribution of Working Groups I, II and III to the Fifth Assessment Report of the Intergovernmental Panel on Climate Change* [Core Writing Team, R.K. Pachauri, and L.A. Meyer (eds.)]. IPCC, Geneva, Switzerland, pp. 117–130.

2 IPCC, 2022, Annex I: Glossary [van Diemen, R., J.B.R. Matthews, V. Möller, J.S. Fuglestvedt, V. Masson-Delmotte, C. Méndez, A. Reisinger, and S. Semenov (eds.)]. In: *IPCC, 2022: Climate Change 2022: Mitigation of Climate Change. Contribution of Working Group III to the Sixth Assessment Report of the Intergovernmental Panel on Climate Change* [P.R. Shukla, J. Skea, R. Slade, A. Al Khourdajie, R. van Diemen, D. McCollum, M. Pathak, S. Some, P. Vyas, R. Fradera, M. Belkacemi, A. Hasija, G. Lisboa, S. Luz, and J. Malley, (eds.)]. Cambridge University Press, Cambridge, UK and New York, doi: 10.1017/9781009157926.020.

3 IPCC, 2022, Annex I: Glossary [van Diemen, R., J.B.R. Matthews, V. Moller, J.S. Fuglestvedt, V. Masson-Delmotte, C. Mendez, A. Reisinger, and S. Semenov (eds.)]. In: *IPCC, 2022: Climate Change 2022: Mitigation of Climate Change. Contribution of Working Group III to the Sixth Assessment Report of the Intergovernmental Panel on Climate Change* [P.R. Shukla, J. Skea, R. Slade, A. Al Khourdajie, R. van Diemen, D. McCollum, M. Pathak, S. Some, P. Vyas, R. Fradera, M. Belkacemi, A. Hasija, G. Lisboa, S. Luz, and J. Malley, (eds.)]. Cambridge University Press, Cambridge, UK and New York, doi: 10.1017/9781009157926.020.

4 IPCC, 2022, Annex I: Glossary [van Diemen, R., J.B.R. Matthews, V. Moller, J.S. Fuglestvedt, V. Masson-Delmotte, C. Mendez, A. Reisinger, and S. Semenov (eds.)]. In: *IPCC, 2022: Climate Change 2022: Mitigation of Climate Change. Contribution of Working Group III to the Sixth Assessment Report of the Intergovernmental Panel on Climate Change* [P.R. Shukla, J. Skea, R. Slade, A. Al Khourdajie, R. van Diemen, D. McCollum, M. Pathak, S. Some, P. Vyas, R. Fradera, M. Belkacemi, A. Hasija, G. Lisboa, S. Luz, and J. Malley, (eds.)]. Cambridge University Press, Cambridge, UK and New York, doi: 10.1017/9781009157926.020.

5 (n.a.), 2016, "What is CCS?" Carbon Capture and Storage Association, www.ccsassociation.org.

6 Qiao, D., Xingjie, L., Kui, Z., Leichuan, T., Guilin, Q., and Jian, Z., 2022, CCS and CCUS Technologies: Giving the Oil and Gas Industry a Green Future, *Frontiers in Energy Research,* 10, https://www.frontiersin.org/articles/10.3389/fenrg.2022.919330, DOI=10.3389/fenrg.2022.919330, ISSN=2296-598X.

7 IPCC, 2022, Annex I: Glossary [van Diemen, R., J.B.R. Matthews, V. Möller, J.S. Fuglestvedt, V. Masson-Delmotte, C. Méndez, A. Reisinger, and S. Semenov (eds.)]. In: *IPCC, 2022: Climate Change 2022: Mitigation of Climate Change. Contribution of Working Group III to the Sixth Assessment Report of the Intergovernmental Panel on Climate Change* [P.R. Shukla, J. Skea, R. Slade, A. Al Khourdajie, R. van Diemen, D. McCollum, M. Pathak, S. Some, P. Vyas, R. Fradera, M. Belkacemi, A. Hasija, G. Lisboa, S. Luz, and J. Malley, (eds.)]. Cambridge University Press, Cambridge, UK and New York, doi: 10.1017/9781009157926.020.

8 IPCC, 2022, Annex I: Glossary [van Diemen, R., J.B.R. Matthews, V. Moller, J.S. Fuglestvedt, V. Masson-Delmotte, C. Mendez, A. Reisinger, and S. Semenov (eds.)]. In: *IPCC, 2022: Climate Change 2022: Mitigation of Climate Change. Contribution of Working Group III to the Sixth Assessment Report of the Intergovernmental Panel on Climate Change* [P.R. Shukla, J. Skea, R. Slade, A. Al Khourdajie, R. van Diemen, D. McCollum, M. Pathak, S. Some, P. Vyas, R. Fradera, M. Belkacemi, A. Hasija, G. Lisboa, S. Luz, and J. Malley, (eds.)]. Cambridge University Press, Cambridge, UK and New York, doi: 10.1017/9781009157926.020.

9 IPCC, 2014, Annex II: Glossary [Mach, K.J., S. Planton, and C. von Stechow (eds.)]. In: *Climate Change 2014: Synthesis Report. Contribution of Working Groups I, II and III to the Fifth Assessment Report of the Intergovernmental Panel on Climate Change* [Core Writing Team, R.K. Pachauri and L.A. Meyer (eds.)]. IPCC, Geneva, Switzerland, pp. 117–130.

10 IPCC, 2022, Annex I: Glossary [van Diemen, R., J.B.R. Matthews, V. Möller, J.S. Fuglestvedt, V. Masson-Delmotte, C. Méndez, A. Reisinger, and S. Semenov (eds.)]. In: *IPCC, 2022: Climate Change 2022: Mitigation of Climate Change. Contribution of Working Group III to the Sixth Assessment Report of the Intergovernmental Panel on Climate Change* [P.R. Shukla, J. Skea, R. Slade, A. Al Khourdajie, R. van Diemen, D. McCollum, M. Pathak, S. Some, P. Vyas, R. Fradera, M. Belkacemi, A. Hasija, G. Lisboa, S. Luz, and J. Malley, (eds.)]. Cambridge University Press, Cambridge, UK and New York, doi: 10.1017/9781009157926.020.

11 IPCC, 2022, Annex I: Glossary [van Diemen, R., J.B.R. Matthews, V. Moller, J.S. Fuglestvedt, V. Masson-Delmotte, C. Mendez, A. Reisinger, and S. Semenov (eds.)]. In: *IPCC, 2022: Climate Change 2022: Mitigation of Climate Change. Contribution of Working Group III to the Sixth Assessment Report of the Intergovernmental Panel on Climate Change* [P.R. Shukla, J. Skea, R. Slade, A. Al Khourdajie, R. van Diemen, D. McCollum, M. Pathak, S. Some, P. Vyas, R. Fradera, M. Belkacemi, A. Hasija, G. Lisboa, S. Luz, and J. Malley, (eds.)]. Cambridge University Press, Cambridge, UK and New York, doi: 10.1017/9781009157926.020.

12 IPCC, 2022, Annex I: Glossary [van Diemen, R., J.B.R. Matthews, V. Moller, J.S. Fuglestvedt, V. Masson-Delmotte, C. Mendez, A. Reisinger, and S. Semenov (eds.)]. In: *IPCC, 2022: Climate Change 2022: Mitigation of Climate Change. Contribution of Working Group III to the Sixth Assessment Report of the Intergovernmental Panel on Climate Change* [P.R. Shukla, J. Skea, R. Slade, A. Al Khourdajie, R. van Diemen, D. McCollum, M. Pathak, S. Some, P. Vyas, R. Fradera, M. Belkacemi, A. Hasija, G. Lisboa, S. Luz, and J. Malley, (eds.)]. Cambridge University Press, Cambridge, UK and New York, doi: 10.1017/9781009157926.020.

13 IPCC, 2022, Annex I: Glossary [van Diemen, R., J.B.R. Matthews, V. Moller, J.S. Fuglestvedt, V. Masson-Delmotte, C. Mendez, A. Reisinger, and S. Semenov (eds.)]. In: *IPCC, 2022: Climate Change 2022: Mitigation of Climate Change. Contribution of Working Group III to the Sixth Assessment Report of the Intergovernmental Panel on Climate Change* [P.R. Shukla, J. Skea, R. Slade, A. Al Khourdajie, R. van Diemen, D. McCollum, M. Pathak, S. Some, P. Vyas, R. Fradera, M. Belkacemi, A. Hasija, G. Lisboa, S. Luz, and J. Malley, (eds.)]. Cambridge University Press, Cambridge, UK and New York, doi: 10.1017/9781009157926.020.

14 (n.a.), February 10, 2020, "CMIP Phase 5 (CMIP5)," World Climate Research Programme, www.wcrp-climate.org/wgcm-cmip/wgcm-cmip5.

15 IPCC, 2022, Annex I: Glossary [van Diemen, R., J.B.R. Matthews, V. Möller, J.S. Fuglestvedt, V. Masson-Delmotte, C. Méndez, A. Reisinger, and S. Semenov (eds.)]. In: *IPCC, 2022: Climate Change 2022: Mitigation of Climate Change. Contribution of Working Group III to the Sixth Assessment Report of the Intergovernmental Panel on Climate Change* [P.R. Shukla, J. Skea, R. Slade, A. Al Khourdajie, R. van Diemen, D. McCollum, M. Pathak, S. Some, P. Vyas, R. Fradera, M. Belkacemi, A. Hasija, G. Lisboa, S. Luz, and J. Malley, (eds.)]. Cambridge University Press, Cambridge, UK and New York, doi: 10.1017/9781009157926.020.

16 UNGA, 2016, "Report of the Open-Ended Intergovernmental Expert Working Group on Indicators and Terminology Relating to Disaster Risk Reduction," United Nations General Assembly (UNGA), 41pp. https://www.preventionweb.net/files/50683_oiewgreportenglish.pdf.

17 (n.a.), 2022, "What is Materiality in the Context of ESG?", enVizi, https://envizi.com/a-guide-to-esg-reporting-frameworks/?_ga=2.162965055.2006466284.1672367224-375724963.1672040420.

18 IPCC, 2022, Annex I: Glossary [van Diemen, R., J.B.R. Matthews, V. Möller, J.S. Fuglestvedt, V. Masson-Delmotte, C. Méndez, A. Reisinger, and S. Semenov (eds.)]. In: *IPCC, 2022: Climate Change 2022: Mitigation of Climate Change. Contribution of Working Group III to the Sixth Assessment*

Report of the Intergovernmental Panel on Climate Change [P.R. Shukla, J. Skea, R. Slade, A. Al Khourdajie, R. van Diemen, D. McCollum, M. Pathak, S. Some, P. Vyas, R. Fradera, M. Belkacemi, A. Hasija, G. Lisboa, S. Luz, and J. Malley, (eds.)]. Cambridge University Press, Cambridge, UK and New York, doi: 10.1017/9781009157926.020.

19 IPCC, 2022, Annex I: Glossary [van Diemen, R., J.B.R. Matthews, V. Moller, J.S. Fuglestvedt, V. Masson-Delmotte, C. Mendez, A. Reisinger, and S. Semenov (eds.)]. In: *IPCC, 2022: Climate Change 2022: Mitigation of Climate Change. Contribution of Working Group III to the Sixth Assessment Report of the Intergovernmental Panel on Climate Change* [P.R. Shukla, J. Skea, R. Slade, A. Al Khourdajie, R. van Diemen, D. McCollum, M. Pathak, S. Some, P. Vyas, R. Fradera, M. Belkacemi, A. Hasija, G. Lisboa, S. Luz, and J. Malley, (eds.)]. Cambridge University Press, Cambridge, UK and New York, doi: 10.1017/9781009157926.020.

20 Allison, C., Stacey, J., Solsbery, L., and Peirce, A., June 2017, "Technical Supplement. The Use of Scenario Analysis in Disclosure of Climate-Related Risks and Opportunities," https://assets.bbhub.io/company/sites/60/2020/10/FINAL-TCFD-Technical-Supplement-062917.pdf.

21 Allison, C., Stacey, J., Solsbery, L., and Peirce, A., June 2017, "Technical Supplement. The Use of Scenario Analysis in Disclosure of Climate-Related Risks and Opportunities," https://assets.bbhub.io/company/sites/60/2020/10/FINAL-TCFD-Technical-Supplement-062917.pdf.

22 Allison, C., Stacey, J., Solsbery, L., and Peirce, A., June 2017, "Technical Supplement. The Use of Scenario Analysis in Disclosure of Climate-Related Risks and Opportunities," https://assets.bbhub.io/company/sites/60/2020/10/FINAL-TCFD-Technical-Supplement-062917.pdf.

23 IPCC, 2022, Annex I: Glossary [van Diemen, R., J.B.R. Matthews, V. Möller, J.S. Fuglestvedt, V. Masson-Delmotte, C. Méndez, A. Reisinger, and S. Semenov (eds.)]. In: *IPCC, 2022: Climate Change 2022: Mitigation of Climate Change. Contribution of Working Group III to the Sixth Assessment Report of the Intergovernmental Panel on Climate Change* [P.R. Shukla, J. Skea, R. Slade, A. Al Khourdajie, R. van Diemen, D. McCollum, M. Pathak, S. Some, P. Vyas, R. Fradera, M. Belkacemi, A. Hasija, G. Lisboa, S. Luz, and J. Malley, (eds.)]. Cambridge University Press, Cambridge, UK and New York, doi: 10.1017/9781009157926.020.

24 IPCC, 2022, Annex I: Glossary [van Diemen, R., J.B.R. Matthews, V. Moller, J.S. Fuglestvedt, V. Masson-Delmotte, C. Mendez, A. Reisinger, and S. Semenov (eds.)]. In: *IPCC, 2022: Climate Change 2022: Mitigation of Climate Change. Contribution of Working Group III to the Sixth Assessment Report of the Intergovernmental Panel on Climate Change* [P.R. Shukla, J. Skea, R. Slade, A. Al Khourdajie, R. van Diemen, D. McCollum, M. Pathak, S. Some, P. Vyas, R. Fradera, M. Belkacemi, A. Hasija, G. Lisboa, S. Luz, and J. Malley, (eds.)]. Cambridge University Press, Cambridge, UK and New York, doi: 10.1017/9781009157926.020.

25 IPCC, 2022, Annex I: Glossary [van Diemen, R., J.B.R. Matthews, V. Moller, J.S. Fuglestvedt, V. Masson-Delmotte, C. Mendez, A. Reisinger, and S. Semenov (eds.)]. In: *IPCC, 2022: Climate Change 2022: Mitigation of Climate Change. Contribution of Working Group III to the Sixth Assessment Report of the Intergovernmental Panel on Climate Change* [P.R. Shukla, J. Skea, R. Slade, A. Al Khourdajie, R. van Diemen, D. McCollum, M. Pathak,

S. Some, P. Vyas, R. Fradera, M. Belkacemi, A. Hasija, G. Lisboa, S. Luz, and J. Malley, (eds.)]. Cambridge University Press, Cambridge, UK and New York, doi: 10.1017/9781009157926.020.

26 Cadbury, Report of the Committee on the Financial Aspects of Corporate Governance, 1992.

27 OECD, G20/OECD Principles of Corporate Governance, 2015.

28 IPCC, 2022, Annex I: Glossary [van Diemen, R., J.B.R. Matthews, V. Möller, J.S. Fuglestvedt, V. Masson-Delmotte, C. Méndez, A. Reisinger, and S. Semenov (eds.)]. In: *IPCC, 2022: Climate Change 2022: Mitigation of Climate Change. Contribution of Working Group III to the Sixth Assessment Report of the Intergovernmental Panel on Climate Change* [P.R. Shukla, J. Skea, R. Slade, A. Al Khourdajie, R. van Diemen, D. McCollum, M. Pathak, S. Some, P. Vyas, R. Fradera, M. Belkacemi, A. Hasija, G. Lisboa, S. Luz, and J. Malley, (eds.)]. Cambridge University Press, Cambridge, UK and New York, doi: 10.1017/9781009157926.020.

29 IPCC, 2022, Annex I: Glossary [van Diemen, R., J.B.R. Matthews, V. Moller, J.S. Fuglestvedt, V. Masson-Delmotte, C. Mendez, A. Reisinger, and S. Semenov (eds.)]. In: *IPCC, 2022: Climate Change 2022: Mitigation of Climate Change. Contribution of Working Group III to the Sixth Assessment Report of the Intergovernmental Panel on Climate Change* [P.R. Shukla, J. Skea, R. Slade, A. Al Khourdajie, R. van Diemen, D. McCollum, M. Pathak, S. Some, P. Vyas, R. Fradera, M. Belkacemi, A. Hasija, G. Lisboa, S. Luz, and J. Malley, (eds.)]. Cambridge University Press, Cambridge, UK and New York, doi: 10.1017/9781009157926.020.

30 Bhatia, P., et al., April 16, 2014, "The Corporate Value Chain (Scope 3) Accounting and Reporting Standard, Supplement to the GHG Protocol Corporate Accounting and Reporting Standard," World Resources Institute (WRI) and World Business Council for Sustainable Development (WBCSD), https://ghgprotocol.org/sites/default/files/standards/Corporate-Value-Chain-Accounting-Reporing-Standard_041613_2.pdf.

31 Ranganathan, J., et al., March 2004, "World Resources Institute (WRI) and World Business Council for Sustainable Development (WBCSD), The Greenhouse Gas Protocol: A Corporate Accounting and Reporting Standard (Revised Edition)," https://ghgprotocol.org/sites/default/files/standards/ghg-protocol-revised.pdf.

32 Bhatia, P., et al., April 16, 2014, "The Corporate Value Chain (Scope 3) Accounting and Reporting Standard, Supplement to the GHG Protocol Corporate Accounting and Reporting Standard," World Resources Institute (WRI) and World Business Council for Sustainable Development (WBCSD), https://ghgprotocol.org/sites/default/files/standards/Corporate-Value-Chain-Accounting-Reporing-Standard_041613_2.pdf.

33 Ranganathan, J., et al., March 2004, "World Resources Institute (WRI) and World Business Council for Sustainable Development (WBCSD), The Greenhouse Gas Protocol: A Corporate Accounting and Reporting Standard (Revised Edition)," https://ghgprotocol.org/sites/default/files/standards/ghg-protocol-revised.pdf.

34 Benson, A., September 14, 2022, "Greenwashing: What Is It?" NerdWallet Compare, Inc., www.nerdwallet.com/article/investing/greenwashing.

35 OHCHR, 2018, "What are Human rights?" Office of the High Commissioner for Human Rights (OHCHR), www.ohchr.org/en/what-are-human-rights.

36 UNESCO, 2018, "Local and Indigenous Knowledge Systems." United Nations Educational, Scientific and Cultural Organization (UNESCO). https://en.unesco.org/links.
37 (n.a.), November 27, 2019, "Regulation (EU) 2019/2088 Of the European Parliament and Of the Council on Sustainability-Related Disclosures in The Financial Services Sector," Official Journal of the European Union, https://eur-lex.europa.eu/legal-content/EN/TXT/PDF/?uri=CELEX:32019R2088.
38 Allison, C., Stacey, J., Solsbery, L., and Peirce, A., June 2017, "Technical Supplement. The Use of Scenario Analysis in Disclosure of Climate-Related Risks and Opportunities," https://assets.bbhub.io/company/sites/60/2020/10/FINAL-TCFD-Technical-Supplement-062917.pdf.
39 IPCC, 2022, Annex I: Glossary [van Diemen, R., J.B.R. Matthews, V. Möller, J.S. Fuglestvedt, V. Masson-Delmotte, C. Méndez, A. Reisinger, and S. Semenov (eds.)]. In: *IPCC, 2022: Climate Change 2022: Mitigation of Climate Change. Contribution of Working Group III to the Sixth Assessment Report of the Intergovernmental Panel on Climate Change* [P.R. Shukla, J. Skea, R. Slade, A. Al Khourdajie, R. van Diemen, D. McCollum, M. Pathak, S. Some, P. Vyas, R. Fradera, M. Belkacemi, A. Hasija, G. Lisboa, S. Luz, and J. Malley, (eds.)]. Cambridge University Press, Cambridge, UK and New York, doi: 10.1017/9781009157926.020.
40 Intergovernmental Panel on Climate Change (IPCC), 2014, Annex II: Glossary [Mach, K.J., S. Planton, and C. von Stechow (eds.)]. In: *Climate Change 2014: Synthesis Report. Contribution of Working Groups I, II and III to the Fifth Assessment Report of the Intergovernmental Panel on Climate Change* [Core Writing Team, R.K. Pachauri and L.A. Meyer (eds.)]. IPCC, Geneva, Switzerland, pp. 117–130, www.ipcc.ch/site/assets/uploads/2018/02/SYR_AR5_FINAL_full.pdf
41 IPCC, 2022, Annex I: Glossary [van Diemen, R., J.B.R. Matthews, V. Möller, J.S. Fuglestvedt, V. Masson-Delmotte, C. Méndez, A. Reisinger, and S. Semenov (eds.)]. In: *IPCC, 2022: Climate Change 2022: Mitigation of Climate Change. Contribution of Working Group III to the Sixth Assessment Report of the Intergovernmental Panel on Climate Change* [P.R. Shukla, J. Skea, R. Slade, A. Al Khourdajie, R. van Diemen, D. McCollum, M. Pathak, S. Some, P. Vyas, R. Fradera, M. Belkacemi, A. Hasija, G. Lisboa, S. Luz, and J. Malley, (eds.)]. Cambridge University Press, Cambridge, UK and New York, doi: 10.1017/9781009157926.020.
42 (n.a.), 2022, "What is Materiality in the Context of ESG?," enVizi, https://envizi.com/a-guide-to-esg-reporting-frameworks/?_ga=2.162965055.2006466284.1672367224-375724963.1672040420.
43 IPCC, 2022, Annex I: Glossary [van Diemen, R., J.B.R. Matthews, V. Möller, J.S. Fuglestvedt, V. Masson-Delmotte, C. Méndez, A. Reisinger, and S. Semenov (eds.)]. In: *IPCC, 2022: Climate Change 2022: Mitigation of Climate Change. Contribution of Working Group III to the Sixth Assessment Report of the Intergovernmental Panel on Climate Change* [P.R. Shukla, J. Skea, R. Slade, A. Al Khourdajie, R. van Diemen, D. McCollum, M. Pathak, S. Some, P. Vyas, R. Fradera, M. Belkacemi, A. Hasija, G. Lisboa, S. Luz, and J. Malley, (eds.)]. Cambridge University Press, Cambridge, UK and New York, doi: 10.1017/9781009157926.020.

44 IPCC, 2022, Annex I: Glossary [van Diemen, R., J.B.R. Matthews, V. Moller, J.S. Fuglestvedt, V. Masson-Delmotte, C. Mendez, A. Reisinger, and S. Semenov (eds.)]. In: *IPCC, 2022: Climate Change 2022: Mitigation of Climate Change. Contribution of Working Group III to the Sixth Assessment Report of the Intergovernmental Panel on Climate Change* [P.R. Shukla, J. Skea, R. Slade, A. Al Khourdajie, R. van Diemen, D. McCollum, M. Pathak, S. Some, P. Vyas, R. Fradera, M. Belkacemi, A. Hasija, G. Lisboa, S. Luz, and J. Malley, (eds.)]. Cambridge University Press, Cambridge, UK and New York, doi: 10.1017/9781009157926.020.

45 IPCC, 2022, Annex I: Glossary [van Diemen, R., J.B.R. Matthews, V. Moller, J.S. Fuglestvedt, V. Masson-Delmotte, C. Mendez, A. Reisinger, and S. Semenov (eds.)]. In: *IPCC, 2022: Climate Change 2022: Mitigation of Climate Change. Contribution of Working Group III to the Sixth Assessment Report of the Intergovernmental Panel on Climate Change* [P.R. Shukla, J. Skea, R. Slade, A. Al Khourdajie, R. van Diemen, D. McCollum, M. Pathak, S. Some, P. Vyas, R. Fradera, M. Belkacemi, A. Hasija, G. Lisboa, S. Luz, and J. Malley, (eds.)]. Cambridge University Press, Cambridge, UK and New York, doi: 10.1017/9781009157926.020.

46 IPCC, 2022, Annex I: Glossary [van Diemen, R., J.B.R. Matthews, V. Moller, J.S. Fuglestvedt, V. Masson-Delmotte, C. Mendez, A. Reisinger, and S. Semenov (eds.)]. In: *IPCC, 2022: Climate Change 2022: Mitigation of Climate Change. Contribution of Working Group III to the Sixth Assessment Report of the Intergovernmental Panel on Climate Change* [P.R. Shukla, J. Skea, R. Slade, A. Al Khourdajie, R. van Diemen, D. McCollum, M. Pathak, S. Some, P. Vyas, R. Fradera, M. Belkacemi, A. Hasija, G. Lisboa, S. Luz, and J. Malley, (eds.)]. Cambridge University Press, Cambridge, UK and New York, doi: 10.1017/9781009157926.020.

47 IPCC, 2022, Annex I: Glossary [van Diemen, R., J.B.R. Matthews, V. Moller, J.S. Fuglestvedt, V. Masson-Delmotte, C. Mendez, A. Reisinger, and S. Semenov (eds.)]. In: *IPCC, 2022: Climate Change 2022: Mitigation of Climate Change. Contribution of Working Group III to the Sixth Assessment Report of the Intergovernmental Panel on Climate Change* [P.R. Shukla, J. Skea, R. Slade, A. Al Khourdajie, R. van Diemen, D. McCollum, M. Pathak, S. Some, P. Vyas, R. Fradera, M. Belkacemi, A. Hasija, G. Lisboa, S. Luz, and J. Malley, (eds.)]. Cambridge University Press, Cambridge, UK and New York, doi: 10.1017/9781009157926.020.

48 IPCC, 2014, Annex II: Glossary [Mach, K.J., S. Planton, and C. von Stechow (eds.)]. In: *Climate Change 2014: Synthesis Report. Contribution of Working Groups I, II and III to the Fifth Assessment Report of the Intergovernmental Panel on Climate Change* [Core Writing Team, R.K. Pachauri and L.A. Meyer (eds.)]. IPCC, Geneva, Switzerland, pp. 117–130.

49 IPCC, 2022, Annex I: Glossary [van Diemen, R., J.B.R. Matthews, V. Möller, J.S. Fuglestvedt, V. Masson-Delmotte, C. Méndez, A. Reisinger, and S. Semenov (eds.)]. In: *IPCC, 2022: Climate Change 2022: Mitigation of Climate Change. Contribution of Working Group III to the Sixth Assessment Report of the Intergovernmental Panel on Climate Change* [P.R. Shukla, J. Skea, R. Slade, A. Al Khourdajie, R. van Diemen, D. McCollum, M. Pathak,

S. Some, P. Vyas, R. Fradera, M. Belkacemi, A. Hasija, G. Lisboa, S. Luz, and J. Malley, (eds.)]. Cambridge University Press, Cambridge, UK and New York, doi: 10.1017/9781009157926.020.

50 IPCC, 2022, Annex I: Glossary [van Diemen, R., J.B.R. Matthews, V. Moller, J.S. Fuglestvedt, V. Masson-Delmotte, C. Mendez, A. Reisinger, and S. Semenov (eds.)]. In: *IPCC, 2022: Climate Change 2022: Mitigation of Climate Change. Contribution of Working Group III to the Sixth Assessment Report of the Intergovernmental Panel on Climate Change* [P.R. Shukla, J. Skea, R. Slade, A. Al Khourdajie, R. van Diemen, D. McCollum, M. Pathak, S. Some, P. Vyas, R. Fradera, M. Belkacemi, A. Hasija, G. Lisboa, S. Luz, and J. Malley, (eds.)]. Cambridge University Press, Cambridge, UK and New York, doi: 10.1017/9781009157926.020.

51 IPCC, 2022, Annex I: Glossary [van Diemen, R., J.B.R. Matthews, V. Moller, J.S. Fuglestvedt, V. Masson-Delmotte, C. Mendez, A. Reisinger, and S. Semenov (eds.)]. In: *IPCC, 2022: Climate Change 2022: Mitigation of Climate Change. Contribution of Working Group III to the Sixth Assessment Report of the Intergovernmental Panel on Climate Change* [P.R. Shukla, J. Skea, R. Slade, A. Al Khourdajie, R. van Diemen, D. McCollum, M. Pathak, S. Some, P. Vyas, R. Fradera, M. Belkacemi, A. Hasija, G. Lisboa, S. Luz, and J. Malley, (eds.)]. Cambridge University Press, Cambridge, UK and New York, doi: 10.1017/9781009157926.020.

52 UNFCCC, 2019, "Land Use, Land-Use Change and Forestry (LULUCF)," United Nations Framework Convention on Climate Change (UNFCCC). https://unfccc.int/topics/land-use/workstreams/land-use--land-use-change-andforestry-lulucf.

53 Allison, C., Stacey, J., Solsbery, L., and Peirce, A., June 2017 "Technical Supplement. The Use of Scenario Analysis in Disclosure of Climate-Related Risks and Opportunities," https://assets.bbhub.io/company/sites/60/2020/10/FINAL-TCFD-Technical-Supplement-062917.pdf.

54 (n.a.), October 2021, "Task Force on Climate-related Financial Disclosures, 2021 Status Report," www.fsb-tcfd.org/publications.

55 Allison, C., Stacey, J., Solsbery, L., and Peirce, A., June 2017 "Technical Supplement. The Use of Scenario Analysis in Disclosure of Climate-Related Risks and Opportunities," https://assets.bbhub.io/company/sites/60/2020/10/FINAL-TCFD-Technical-Supplement-062917.pdf.

56 Allison, C., Stacey, J., Solsbery, L., and Peirce, A., June 2017 "Technical Supplement. The Use of Scenario Analysis in Disclosure of Climate-Related Risks and Opportunities," https://assets.bbhub.io/company/sites/60/2020/10/FINAL-TCFD-Technical-Supplement-062917.pdf.

57 IPCC, 2022, Annex I: Glossary [van Diemen, R., J.B.R. Matthews, V. Möller, J.S. Fuglestvedt, V. Masson-Delmotte, C. Méndez, A. Reisinger, and S. Semenov (eds.)]. In: *IPCC, 2022: Climate Change 2022: Mitigation of Climate Change. Contribution of Working Group III to the Sixth Assessment Report of the Intergovernmental Panel on Climate Change* [P.R. Shukla, J. Skea, R. Slade, A. Al Khourdajie, R. van Diemen, D. McCollum, M. Pathak, S. Some, P. Vyas, R. Fradera, M. Belkacemi, A. Hasija, G. Lisboa, S. Luz, and J. Malley, (eds.)]. Cambridge University Press, Cambridge, UK and New York, doi: 10.1017/9781009157926.020.

58 IPCC, 2022, Annex I: Glossary [van Diemen, R., J.B.R. Matthews, V. Moller, J.S. Fuglestvedt, V. Masson-Delmotte, C. Mendez, A. Reisinger, and S. Semenov (eds.)]. In: *IPCC, 2022: Climate Change 2022: Mitigation of Climate Change. Contribution of Working Group III to the Sixth Assessment Report of the Intergovernmental Panel on Climate Change* [P.R. Shukla, J. Skea, R. Slade, A. Al Khourdajie, R. van Diemen, D. McCollum, M. Pathak, S. Some, P. Vyas, R. Fradera, M. Belkacemi, A. Hasija, G. Lisboa, S. Luz, and J. Malley, (eds.)]. Cambridge University Press, Cambridge, UK and New York, doi: 10.1017/9781009157926.020.
59 UNDESA, 2018, "World Urbanization Prospects, the 2018 Revision," Technical Report. United Nations Department of Economic and Social Affairs (UNDESA), New York, NY, USA.
60 WCED, 1987, *Our Common Future*, World Commission on Environment and Development (WCED), Oxford University Press, Oxford, UK.
61 IPCC, 2022, Annex I: Glossary [van Diemen, R., J.B.R. Matthews, V. Möller, J.S. Fuglestvedt, V. Masson-Delmotte, C. Méndez, A. Reisinger, and S. Semenov (eds.)]. In: *IPCC, 2022: Climate Change 2022: Mitigation of Climate Change. Contribution of Working Group III to the Sixth Assessment Report of the Intergovernmental Panel on Climate Change* [P.R. Shukla, J. Skea, R. Slade, A. Al Khourdajie, R. van Diemen, D. McCollum, M. Pathak, S. Some, P. Vyas, R. Fradera, M. Belkacemi, A. Hasija, G. Lisboa, S. Luz, and J. Malley, (eds.)]. Cambridge University Press, Cambridge, UK and New York, doi: 10.1017/9781009157926.020.
62 (n.a.), (n.d.), "Sustainable Finance Disclosure Regulation (SFDR)," Environmental Social Governance Solutions, https://esgsolutionsltd.com/portfolio-item/sustainable-finance-disclosure-regulation-sfdr.
63 (n.a.), 2022, "Transition Risk Report," GRESB, www.gresb.com/nl-en/products/transition-risk-tool.
64 UNFCCC, 1992, "United Nations Framework Convention on Climate Change," United Nations Framework Convention on Climate Change (UNFCCC), https://unfccc.int/resource/docs/convkp/conveng.pdf.
65 Allison, C., Stacey, J., Solsbery, L., and Peirce, A., June 2017, "Technical Supplement. The Use of Scenario Analysis in Disclosure of Climate-Related Risks and Opportunities," https://assets.bbhub.io/company/sites/60/2020/10/FINAL-TCFD-Technical-Supplement-062917.pdf.
66 IPCC, 2022, Annex I: Glossary [van Diemen, R., J.B.R. Matthews, V. Möller, J.S. Fuglestvedt, V. Masson-Delmotte, C. Méndez, A. Reisinger, and S. Semenov (eds.)]. In: *IPCC, 2022: Climate Change 2022: Mitigation of Climate Change. Contribution of Working Group III to the Sixth Assessment Report of the Intergovernmental Panel on Climate Change* [P.R. Shukla, J. Skea, R. Slade, A. Al Khourdajie, R. van Diemen, D. McCollum, M. Pathak, S. Some, P. Vyas, R. Fradera, M. Belkacemi, A. Hasija, G. Lisboa, S. Luz, and J. Malley, (eds.)]. Cambridge University Press, Cambridge, UK and New York, doi: 10.1017/9781009157926.020.
67 IPCC, 2022, Annex I: Glossary [van Diemen, R., J.B.R. Matthews, V. Moller, J.S. Fuglestvedt, V. Masson-Delmotte, C. Mendez, A. Reisinger, and S. Semenov (eds.)]. In: *IPCC, 2022: Climate Change 2022: Mitigation of Climate Change. Contribution of Working Group III to the Sixth Assessment*

Report of the Intergovernmental Panel on Climate Change [P.R. Shukla, J. Skea, R. Slade, A. Al Khourdajie, R. van Diemen, D. McCollum, M. Pathak, S. Some, P. Vyas, R. Fradera, M. Belkacemi, A. Hasija, G. Lisboa, S. Luz, and J. Malley, (eds.)]. Cambridge University Press, Cambridge, UK and New York, doi: 10.1017/9781009157926.020.

Appendix 4

ESG Acronyms

Acronym	*Meaning*
1.5DS	1.5-Degree Celsius (1.5°C) Target
2DS	2-Degree Celsius (2°C) Target
AIFMs	Alternative Investment Fund Managers
AIGCC	Asia Investor Group on Climate Change (Asia)
AUM	Assets under Management
BEIS	Business, Energy, and Industrial Strategy
BREEAM	Building Research Establishment Environmental Assessment Methodology
CA100+	Climate Action 100+
CCIV	Corporate Collective Investment Vehicles
CCS	Carbon Capture and Sequestration/Storage
CCUS	Carbon Capture, Utilization, and Storage
CDSB	Climate Disclosure Standards Board
CDD	Cooling Degree Day
CDP	Carbon Disclosure Project
CDSB	Climate Disclosure Standards Board
CH_4	Methane
CII	Climate Innovation Index
CO_2	Carbon Dioxide
COP21	Conference of Parties
CMIP5	Coupled Model Intercomparison Project 5
CR	Climate Registry
CRD	Corporate Reporting Dialogue
CRE	Commercial Real Estate
CRREM	Carbon Risk Real Estate Monitor
CSR	Corporate Social Responsibility
CSRD	Corporate Sustainability Reporting Directive
CVaR	Carbon Value at Risk
DNSH	Do Not Significantly Harm
DSJI	Dow Jones Sustainability Index

(*Continued*)

(Continued)

Acronym	*Meaning*
EBA	European Banking Authority
EBITDA	Earnings Before Interest, Taxes, Depreciation, and Amortization
EBRD	European Bank for Reconstruction and Development
EDTF	Enhanced Disclosure Task Force
EEDI	Energy Efficiency Design Index
EF	Emission Factor
EfW	Energy from Waste
EICC	Electronic Industry Citizenship Coalition
EIOPA	European Insurance and Occupational Pensions Authority
ELTIFs	European Long-Term Investment Funds
EPC	Energy Performance Certificate
EPIC	Embankment Project for Inclusive Capitalism
EPRA	European Public Real Estate Association
ESA	European Supervisory Authorities
ESG	Environmental, Social, and Governance
ESMA	European Securities and Markets Authority
ESMS	Environmental and Social Management System
ETF	Exchange-Traded Fund
EFRAG	European Financial Reporting Advisory Group
EuSEF	European Social Entrepreneurship Funds
EuVECA	European Venture Capital Funds
EVIC	Enterprise Value Including Cash
FCA	Financial Conduct Authority
FLO	Fairtrade Labelling Organization
FMP	Financial Market Participants
FSB	Financial Stability Board
FSO	Forest Stewardship Council
FTSE	Financial Times Stock Exchange
GCM	General Circulation Models
G4	GRI's Reporting Framework
G20	Group of Twenty
G30	Group of Thirty
GBP	Green Bonds Principles
GDR	Global Depositary Receipt
GHG	Greenhouse Gas
GIC	Global Investor Coalition on Climate Change
GRESB	Global Real Estate Sustainability Benchmark
GRI	Global Reporting Initiative
GSAM	Goldman Sachs Asset Management
GWP	Global Warming Potential

(Continued)

(*Continued*)

Acronym	*Meaning*
HFCs	Hydrofluorocarbons
HSI	Hang Seng Corporate Sustainability Index
HVAC	Heating, Ventilation and Air Conditioning
IAG	Investor Advisory Group
IAM	Integrated Assessment Models
IASB	International Accounting Standards Board
IBIPs	Insurance-Based Investment Products
ICBC	Industrial and Commerce Bank of China
ICMA	International Capital Markets Association
IEA	International Energy Agency
IFRS	International Financial Reporting Standards
IGCC	Investor Group on Climate Change (Australia and New Zealand)
IIGCC	Institutional Investors Group on Climate Change (Europe)
IIRC	International Integrated Reporting Council
INCR	Investor Network on Climate Risk (North America)
INDC	Intended Nationally Determined Contribution
INREV	European Investors in Non-Listed Real Estate
IORPs	Institutions for Occupational Retirement Provision
IOSCO	International Organization of Securities Commissions
IPCC	Intergovernmental Panel on Climate Change
IPMS	International Property Measurement Standards
ISAE 3000	Assurance Statements for Sustainability Disclosures
ISAE	International Standards on Assurance Engagements
ISSB	International Sustainability Standards Board
ISO 14064	International Standards for GHG Emissions Measurement
ISO 26000	International Standards for Social Responsibility
ITR	Implied Temperature Rise
LEED	Leadership in Energy and Environmental Design
LFGTE	Landfill-gas-to-energy
LTIP	Long-Term Incentive Plan
LULUCF	Land Use/Land Use Change/Forestry
MiFID II	Markets in Financial Instruments Directive 2
MRV	Monitoring, Reporting, and Verification
MSW	Municipal Solid Waste
N_2O	Nitrous Oxide
NBFI	Nonbank Financial Institution
NCSSM	National Commission on Securities and Stock Market
NDC	Nationally Determined Contributions
NFR Directive	Non-Financial Reporting Directive
NGFS	Network for Greening the Financial System

(*Continued*)

(Continued)

Acronym	*Meaning*
NGO	Non-Governmental Organization
NZE2050	Net Zero Emissions by 2050
OECD	Organisation for Economic Co-operation and Development
OMFIF	Official Monetary and Financial Institutions Forum
OMZ	Oxygen Minimum Zone
PAI	Principal Adverse Impacts
PCAF	Partnership for Carbon Accounting Financials
PEPP	Providers of Pan-European Personal Pension Products
PFCs	Perfluorocarbons
PPA	Power Purchase Agreement
PRI	Principles for Responsible Investment
PSEG	Public Service Enterprise Group
RBC	Responsible Business Conduct
REIT	Real Estate Investment Trust
RTS	Regulatory Technical Standards
sBPR	Sustainability Best Practice Recommendations
SAM	Sustainable Asset Management
SASB	Sustainability Accounting Standards Board
SBT	Science Based Targets
SBTi's	Science Based Targets Initiative's
SBP	Social Bonds Principles
SCM	Stakeholder Capitalism Metrics
SDA	Sectoral Decarbonisation Approach
SDGs	Sustainable Development Goals
SF_6	Sulfur Hexafluoride
SFDR	Sustainable Finance Disclosure Regulation
SLB	Sustainability-Linked Bonds
SLL	Sustainability-Linked Loans
SPTs	Sustainability Performance Targets
SSCF	Sustainable Supply Chain Finance
STC	SAS Trustee Corporation
TCFD	Task Force on Climate-related Financial Disclosures
TFEU	Treaty on the Functioning of the European Union
TSC	The Sustainability Consortium
UCITS	Collective Investment In Transferable Securities
UNDP	United Nations Development Programme
UNEP FI	United Nations Environment Programme Finance Initiative
UNFCCC	United Nations Framework Convention on Climate Change
UNGC	United Nations Global Compact
UNPRI	United Nations Principles for Responsible Investment

(Continued)

(Continued)

Acronym	*Meaning*
WACI	Weighted Average Carbon Intensity
WBCSD	World Business Council for Sustainable Development
WGBC	World Green Building Council
WRI	World Resources Institute
WTE	Waste-to-Energy

Index

Note: **Bold** page numbers refer to tables and *italic* page numbers refer to figures.

For Product Safety Concerns and Information please contact our EU
representative GPSR@taylorandfrancis.com
Taylor & Francis Verlag GmbH, Kaufingerstraße 24, 80331 München, Germany

www.ingramcontent.com/pod-product-compliance
Ingram Content Group UK Ltd.
Pitfield, Milton Keynes, MK11 3LW, UK
UKHW020204200926
475702UK00004B/609

* 9 7 8 1 0 3 2 6 7 2 4 8 9 *